Lost Wonders

Vanished Creatures of Aotearoa

Lost Wonders

Sarah Ell

ALLEN&UNWIN
SYDNEY • MELBOURNE • AUCKLAND • LONDON

First published in 2020

Allen & Unwin
Level 2, 10 College Hill, Freemans Bay
Auckland 1011, New Zealand
Phone: (64 9) 377 3800
Email: auckland@allenandunwin.com
Web: www.allenandunwin.co.nz

83 Alexander Street
Crows Nest NSW 2065, Australia
Phone: (61 2) 8425 0100

A catalogue record for this book is available from the National Library of New Zealand.

ISBN 978 1 98854 744 2

Design by Megan van Staden
Set in Tiempos and Futura
Printed and bound in Australia by Pegasus Media and Logistics

10 9 8 7 6 5 4 3 2

The paper in this book is FSC® certified. FSC® promotes environmentally responsible, socially beneficial and economically viable management of the world's forests.

For my children, Florian and Natalie—
may you live in a world where no more
of the wonders of Aotearoa are lost

Contents

Part 2: Lost and found

Part 3: Almost lost?

Introduction:

The opening of the treasure box

Imagine that you have travelled back in time to the Aotearoa New Zealand of a thousand years ago. It is a land of vast forests. The three main islands and the hundreds of other small ones around the coastline are almost entirely clothed in rich, verdant plant life, from the subtropical jungles of the north to the gnarled and moss-covered beech and rimu of the south. What we know today as the dry, treeless Canterbury Plains (Kā Pākihi Whakatekateka a Waitaha) are covered with a thick forest of tōtara, kahikatea and matai. Only the tops of the highest mountains and the coldest interior regions of what we now call Central Otago are not cloaked in trees. The North Island (Te Ika-a-Māui) is one gigantic forest, from the seashore all the way into its heart.

This land of forests is also a land of birds. The only mammals present are the seals and sea lions, which loll and roar on the rocky coastline, and the secretive bats, which emerge at dusk to catch insects in flight. In the absence of other animals, hundreds of unique bird species have evolved to fill every ecological niche. Without any furry four-footed predators to fear, many of these birds have become flightless; their wings are stumpy and useless, persisting for show only. The birds are loud and curious. They have never seen a human being and have no fear. They are lords of the forest.

People often say that Aotearoa is a young country, but its roots go further back than we realise. Scientists believe that the islands we live on are just the tips of huge mountains that rise from a much larger undersea continent, Zealandia, 94 per cent of which lies under the Pacific Ocean. This continent, nearly 5 million square kilometres in area, stretches from New Caledonia in the north down to the islands of Aotearoa and beyond, to the south and east, and was once part of the southern supercontinent Gondwana. Zealandia split apart from this landmass about 80 million years ago. As the Tasman Sea opened up, much of Zealandia began to slowly sink below the waves.

Today, however, the islands we know as New Zealand are gradually rising from the sea again. Pressure between the Australian and Pacific tectonic plates (sections of the Earth's crust) pushes up the spine of the South Island (Te Waipounamu or Te Waka a Māui) along the Southern Alps / Kā Tiritiri o te Moana. This is the scientific explanation behind the Māori legend of Māui in his waka dragging up a giant fish from the ocean depths for humans to live on.

What all this geological toing and froing meant for the unique plants and animals of Aotearoa was a complete isolation in which they could evolve. In 1990, British naturalist and television presenter David Bellamy dubbed New Zealand 'Moa's Ark', an apt description for this floating piece of land and its precious cargo. Another description might be a waka huia, a treasure box, containing a cache of wonderful things. Giant parrots which lived on the ground. A huge bird, taller than a human, which reached out its long neck to browse leaves like a giraffe. A nocturnal worm-hunter with nostrils at the end of its beak.

Over time, other species arrived here by accident, blown across the Tasman or out of the Pacific by boisterous winds, but the majority of the species which developed here seem likely to have 'caught the boat' when Zealandia drifted away from Gondwana. Surrounded by thousands of square kilometres of sea, tucked away in the windy southern latitudes far below the populated tropical islands of the Pacific, and on the other side of the world from the centres of Western civilisation, Aotearoa New Zealand slumbered in splendid isolation. It was the last significant landmass in the world to remain undiscovered and uninhabited.

Then, around 800 years ago, the first humans arrived. It could have been just one waka, maybe more—no one knows for sure. But the treasure box had been found and, once found, it was opened. As the Polynesian people settled these islands, they discovered the wondrous birds and life-giving trees, the abundant seafood and the precious materials such as pounamu (greenstone) and matā (obsidian). They utilised the resources they found to survive, taking only what they needed for food, construction, tools and decoration.

Although there were few people living in Aotearoa relative to its size—even at the time of Captain James Cook's first expedition in 1769–70 there were still only around 100,000 Māori living here—their presence had a fatal impact on the pristine environment. Māori hunted the larger birds for food, skins and feathers, and with human settlement came fire. The deforestation that would see nearly three-quarters of New Zealand's native tree cover destroyed forever began as early as the thirteenth and fourteenth centuries. Large tracts of bush around the eastern coasts and, most notably, the Canterbury Plains were burned off, to make it easier to hunt large prey such as moa and to create land for agriculture. The fires destroyed not only many precious plants but also the habitats of many birds, insects and smaller creatures.

The first Polynesian settlers also brought with them two strange beasts that the birds of Aotearoa had not encountered before: kurī, the Polynesian dog, and kiore, the Pacific rat. Jumping ashore as waka grounded on empty beaches, they shook off their wet fur and set out to explore. It was the beginning of the end for many of the wonderful creatures aboard the moa's ark.

While nowhere near as destructive as the pest animals the Europeans would later bring, these four-legged predators ran riot in the untouched forests of Tāne, killing adult birds and eating their eggs and young. The rats preyed on native frogs and lizards, too. By the time the first organised European settlements were underway in 1840, it is estimated that 6.7 million hectares of forest had been destroyed and at least 30 species of bird had already been hunted, preyed upon and marginalised into oblivion.

Once European settlers arrived, the population of New Zealand grew—and it grew fast. As more people began to live here, bringing with them strange animal predators, the death warrants of many more species were effectively signed. In the early 1800s, trading and whale ships started visiting our shores, taking trees and decimating the populations of marine mammals in our waters. Then the Treaty of Waitangi was signed in 1840, and the destruction began in earnest. The people who came to live in the newly declared British colony needed food and shelter. They killed birds to eat, but far greater damage was done by habitat destruction.

The new settlers chopped down forests to build houses and ships, and to clear land to farm animals and grow food. Like the Polynesians before them, they brought pest animals that wreaked havoc on the defenceless birds of the treasure box: rats, larger and more tenacious than the kiore, which could swim and climb; dogs and cats, which hunted for sport; browsing animals like possums and deer; then stoats, ferrets and weasels, introduced to control rabbits but which then ran amok too, and continue to do so. These two things did the damage: the devastation of the forests and the introduction of rapacious pests against which Aotearoa's birds had no defences.

Today, the actions of conservationists and government has slowed and all but stopped the felling of native forest, and steps have been taken to preserve what remains. But we are still fighting an epic battle against the persistent predators introduced by European settlers—they present the most serious threat to the survival of our unique native species. Whether or not we win the battle against them remains to be seen.

Over the billions of years that life on Earth has been evolving, species have come and gone. On one level, extinction is a natural process—if it didn't exist, we'd still be sharing the Earth with dinosaurs (or we might not have evolved into existence at all!). New Zealand's coastlines would still be populated by giant penguins and pelicans with teeth. But most of this book is about man-made extinctions—the extermination of species through human greed, stupidity or simple carelessness.

This process of extinction is not natural. It is tragic, and in many cases it is, or should have been, avoidable. Once a species has become extinct—when the last specimen that carries the particular code of DNA that makes it special and unique, distinct from all other life on Earth, has been extinguished—then it is gone forever.

Scientists talk of 'waves' of extinction which have occurred throughout the Earth's history. The last major wave, when massive environmental change caused the widespread extinction of a large number of species, came at the end of the age of the dinosaurs, some 65 million years ago. But some scientists believe we are now experiencing the sixth great wave of extinction. The natural average of species becoming extinct is one every 1000 years, but it is estimated that now, somewhere in the world, at least one species is becoming extinct *every year*.

The process of species being lost forever accelerated in the 1800s, as a rapidly expanding human population spread around the planet, cutting down forests, building cities, dumping waste and otherwise destroying the natural environment. Some species were hunted to their doom, such as the once-abundant passenger pigeon of the American West, while others were eradicated by carelessness, when critical habitat was destroyed. Sometimes people have fought to save a species from extinction, but more often they did not notice until it was too late. By then, something unique and precious had gone from the world. Now, we also face climate change, which may have great impacts on species that are already struggling.

A Global Assessment Report on Biodiversity and Ecosystem Services presented to the United Nations in May 2019 concluded that one million species worldwide are currently at risk of extinction due to human activity. Under the New Zealand Threat Classification System (NZTCS), as of mid 2019, 23 of our bird species are 'nationally critical' and face an immediate high risk of extinction. This category includes the kākāpō and black robin, whose stories feature in this book. A further fourteen bird species, including the iconic kea, are considered nationally endangered and are at high risk of extinction in the short term; and another 33 species are nationally vulnerable and at risk of extinction in the medium term, including many of the rare species of the Chatham Islands and the takahē, which you will also read about in this book. And that is just birds—there are many hundreds of species of marine mammals and fish, insects, reptiles, amphibians and invertebrates which are at serious risk of extinction.

Since humans first arrived in Aotearoa, at least 53 species of birds and many more smaller creatures and plants have become extinct. About 15 of those have been lost since the Treaty of Waitangi was signed in 1840. We can never recover the lost wonders of Aotearoa New Zealand, those gems from the treasure box which have passed from human sight for eternity. We can only remember the species now gone from our world, and wonder at what richness and beauty they once brought to it.

The aim of this book is to share with you the stories of our lost species and to tell you how they came to their sad ends—not to make you feel bad about what happened or lay blame, but so that we can learn from experience. We can't bring our lost species back, but we can pay tribute to them and learn from their extinction in order to not make the same mistakes again. It also tells the stories of a lucky few which have been brought back from the brink—or, in the case of species such as the takahē and tāiko, apparently back from the dead—and some of those still teetering on the edge of it.

Previous generations cannot and should not be judged too harshly for what they did and the part they played in these extinctions. We cannot expect those who lived in the past to hold the same values that we hold today and to share our beliefs about the preciousness of living things. To them, the natural world was a commodity to be exploited; for survival, sometimes, but often for profit.

It is up to us, today's generations, to make sure that the tide of extinctions is stemmed. For some species it may already be too late, but we must do what we can. Every time a species disappears forever from the world, we are the poorer for it. We cannot let this continue to happen.

—Sarah Ell, 2020

Part 1
The lost

Mosasaur

1. What came before: New Zealand's dinosaurs

Joan Wiffen got out of the car and took a deep breath of fresh back-country air. It was a relief after the stuffy warmth of the car. They had travelled for hours to get here, into the hills behind central Hawke's Bay. This was remote, rugged country, and they'd got lost several times along the way. Joan's husband, Pont, had driven along the windy, dusty gravel roads through cut-over native bush. Her son, Chris, had been folded into the back of the car along with the tent and provisions, and Joan had navigated, poring over an unclear map. By the time they'd reached the bridge over the Mangahouanga Stream that they were looking for, it was late afternoon and, even though it was early December, the air was cooling in the deep valley.

Joan took the map out of her parka pocket, unfolded it and looked again at the tiny words inked on to it: 'Reptilian bones in beds of brackish water'. They had to be here somewhere! She felt excitement rising in her chest as she imagined what lay in the rocks below. What had started out as a hobby—looking for fossils (ancient bones turned into rock), signs of New Zealand's distant past—had become almost an obsession. It was an unusual interest for a farmer's wife in the 1970s, but Joan didn't care.

'I'm going down,' she called to Pont and Chris.

She started to clamber down the steep side of the gully below the road bridge, forcing her way through damp fern and struggling to keep her balance on the moss-covered rocks. Reaching the stream bed, she looked around—and, suddenly, there they were. Once her eyes adjusted to the shadows, she could see fossils wherever she looked. It was as if every boulder in the river contained the imprint of some long-lost life form.

Chris and Pont joined her, wading into the fast-flowing cold water to take a closer look. They saw ancient shark teeth, the skeletons of fish and the imprint of scales, the shapes of shells and curly sea creatures.

The gully rang with cries of 'Here!' and 'Look at this!' and 'I've found an ammonite!' The Wiffens waded happily for hours, turning over rock after rock and finding treasure after treasure.

There was no doubt that the family had uncovered a rich site containing some of the most unusual and scientifically valuable fossils yet found in this country. But the greatest treasure of all lay hidden for the moment. Joan didn't know it yet, but before too long she would uncover the remains of New Zealand's first dinosaur.

Parts of New Zealand are very ancient indeed. What we know now as the islands of Aotearoa may have risen from the sea only 20 million or so years ago, but the rocks underneath the thin covering of soil have been around for much, much longer.

Scientists believe that these islands we know as Aotearoa New Zealand are all that remains visible above the surface of a massive undersea continent called Zealandia. The bedrock that forms the skeleton of New Zealand today once lay on the fringes of a larger landmass called Gondwana, which broke up to create the continents of Australia, Antarctica, Africa and South America, and the subcontinent of India. The oldest rocks found in New Zealand are around 500 million years old. They are formed from sediments which ran off the edge of Gondwana and created new rock.

About 80 million years ago, forces within the

Earth caused Zealandia to split away from the landmasses which would become Australia and Antarctica. A rift opened between the two chunks, and the sea flooded in. Zealandia was afloat; a drifting continent about half the size of modern Australia. As it floated away, it carried a cargo of plants and animal life that would thereafter develop in isolation, separated from all the other continents of the world by thousands of square kilometres of ocean.

Over time, the continental crust cooled and became denser, and the majority of Zealandia sank beneath the waves. Around 25–35 million years ago, only a third of the land we live on today was visible. Then, a long, slow period of a process called 'uplift' began. Two great tectonic plates—the Australian and the Pacific—started pushing and rubbing against each other, forcing mountains to rise from the sea. Volcanoes sprung up along the line where the two plates met, creating new areas of land in what would become the North Island. The Earth's internal forces became Māui, pulling his fish out of the sea.

The fossil record shows that this land was populated by birds, reptiles such as freshwater crocodiles, and even some small mammals—bats, like the ones which still live here today, as well as an animal more like a mouse, known as the 'St Bathans mammal' after the town near which its remains were found in Central Otago. There were also tuatara-like lizards and, in the seas surrounding the islands, marine mammals such as seals and whales as well as gigantic sharks. The unique flora and fauna of Aotearoa evolved and diversified untouched by outside influences until the arrival of the first human beings in the very last blip of geological time, around 800 years ago.

Before the majority of Zealandia sank below the waves, leaving just the islands of Aotearoa sticking up out of the sea, it was home to land and flying dinosaurs and the seas surrounding it were populated by primitive reptiles and amphibians. It wasn't long after European settlement that evidence of their presence was found by early geologists.

Europeans settled in Aotearoa not long after the identification of the first-ever dinosaur fossils, in England. One of the earliest dinosaurs to be identified and named was found by doctor Gideon Mantell and his wife, Mary Ann (whose son Walter will feature later in this book). In 1822 the Mantells found several large fossilised teeth on a roadside in Sussex, England, that Gideon decided must have come from some kind of ancient lizard, which he called an iguanodon. (After Gideon's death, one of these teeth was brought to New Zealand by Walter, and it now resides in the Te Papa museum in Wellington.) Based on finds like the Mantells', anatomist Richard Owen coined the term dinosaur, meaning 'fearfully great lizard', in 1841.

Around this time, a large variety of fossils—

the remains of strange creatures that seemed to have walked the Earth long ago but which had now disappeared—were discovered. This led to increasing scientific and public discussion about the idea of extinction. Traditional Christian belief said that God created the world and everything in it over a period of six days. How, then, could there be creatures that had disappeared? Had God put fossils in the rocks? Or was the truth much more complicated than previously thought, involving a much longer period of time and a gradual process of changing life on Earth—a theory which would become known as evolution.

European settlement of New Zealand coincided with these early days of palaeontology, when scientists and collectors were beginning to delve into the mysteries of the history of life on Earth. While investigating what sorts of rocks were found here and looking for valuable minerals and resources, the new arrivals in Aotearoa found other things as well. They were intrigued to find evidence of past life forms buried in the rocks of this distant group of islands.

The first New Zealand fossils officially found and mentioned in scientific literature were a collection of marine reptile bones from Waipara in North Canterbury (Waitaha), discovered by Thomas Cockburn-Hood in 1859. Cockburn-Hood sent the bones to Richard Owen in London in the United Kingdom, who declared in 1861 that they were from a previously undiscovered species of plesiosaur. Cockburn-Hood sent another large shipment of marine reptile bones to England in 1869, but these were all lost when the ship *Matoaka*, which was also carrying a considerable amount of gold bullion, disappeared in the Southern Ocean after sailing from Lyttelton, near Christchurch (Ōtautahi). Other early finds, from Waipara and Haumuri Bluffs, on the South Island coast south of Kaikōura, also went astray. Thirty-nine cases of fossil reptile bones were sent to American expert Edward Drinker Cope in 1890, but he died before examining them; later investigations could find no trace of them.

Much of the investigation of New Zealand's rocks in the nineteenth century was not to uncover signs of ancient life forms, but to identify resources which people could exploit. Fossil wood—that is, coal—was particularly valuable at the time, because machines were powered by fuel-hungry steam engines. Limestone, made up of the fossilised bodies of millions of marine vertebrates, could be ground up for fertiliser, or quarried for building. Certain rock types were known to hold gold and other precious minerals. Early geologists were looking for things that humans could use.

However, Julius von Haast, a Prussian scientist who was the provincial geologist for Canterbury in the 1860s, made it his mission to establish a natural-history museum in Christchurch, and he actively sought fossil specimens. As well as the many moa bones he collected and used to trade with museums worldwide (see chapter 3), he also employed a fossil collector, Alexander McKay.

McKay came to New Zealand from Scotland in 1863, initially as a gold prospector. He became Haast's geological man in the field, travelling widely around the country and collecting around 120,000 fossils in the late 1800s. He also collected fossils for the Colonial Museum in Wellington, the forerunner of today's Te Papa. However, although many fossils of marine reptiles and other smaller species were found, no evidence was ever discovered that indicated that true dinosaurs—reptiles which lived on land and walked with their hind legs vertically under their hips, not splayed out like a crocodile—ever inhabited New Zealand. Until, that is, a farmer's wife called Joan Wiffen started poking around in inland Hawke's Bay in the 1970s.

The first half of the twentieth century was a quiet time for fossil discovery in New Zealand. While geologists continued to look for natural resources with financial value, such as coal and oil, fossil collecting was mainly left to amateurs and enthusiasts. Worldwide, interest in 'popular' prehistoric animals like the mighty dinosaurs had faded among serious palaeontologists. Dozens of scientific papers on New Zealand's extinct marine reptiles were published from 1860 to 1894, but not a single one was produced for another 80 years.

Joan Wiffen didn't know that 'real' scientists believed no land dinosaurs had ever lived in New Zealand, so she wasn't put off looking for them. Joan and Pont lived on a farm in Haumoana, south of Napier, and became interested in geology after attending a night class at the nearby school. They joined a rocks and minerals club, and travelled to Australia to view fossils there. Joan became determined to find significant fossils in New Zealand, and she tracked down a likely site on an old survey map which indicated the presence of fossilised reptile bones in a riverbed in the back of northern Hawke's Bay. In December 1972 the family mounted an expedition there and discovered what would become known as 'the Valley of the Dragons'.

Using textbooks, Joan taught herself to identify the bones they extracted from the fossil-bearing rocks with explosives, grinders and acid baths. The Wiffens found mainly marine reptile fossils at first. It was not until 1979 that American dinosaur expert Dr Ralph Molnar identified an unusual brown bone that Joan had uncovered as a tailbone from a theropod dinosaur—an upright, walking dinosaur which would have lived on land. She had found astonishing proof that, just as in many other parts of the world, dinosaurs had roamed the portion of the Earth which became New Zealand.

While Dr Molnar was the first to present this find to the world, Joan soon published her own scientific papers on her amazing finds. Despite her lack of formal training, she became one of the foremost experts on reptiles from the Cretaceous Period. Already 50 years old at the time of her great discovery, she enjoyed a second career, travelling around the world sharing her knowledge. She continued to make incredible finds at Mangahouanga, including bones from a flying pterosaur and, later in her life, from a titanosaur, the largest type of dinosaur to ever live.

The scientific community was initially wary of the findings of an amateur—and a woman at that—but Joan's contribution became respected worldwide, and she was awarded an honorary doctorate by Massey University in 1994. She also pushed for the Mangahouanga site to be protected by a Queen Elizabeth II National Trust open-space covenant, to preserve the site for fossil-collecting and research.

Thanks to Wiffen's work, and to later discoveries by other palaeontologists, we now know that Aotearoa was formerly home to at least a dozen different kinds of land dinosaurs. Just one has been found from the Jurassic Period: a small theropod which would have been about 1 metre tall, identified from a single toe bone found at Port Waikato. From the later Cretaceous Period—the end of the reign of the dinosaurs—scientists know of at least three carnivores, ranging in size from 2 metres to maybe 10 metres long. There were also three types of herbivores: an armoured ankylosaur, a hypsilophodont which ran on its hind legs, and a large sauropod, with a long neck and tail. In the skies of the late Cretaceous there were pterosaurs, and in the seas around the islands were swimming reptiles such as plesiosaurs and mosasaurs, turtles and squid-like belemnites and ammonites.

No one knows why the dinosaurs became extinct. After millions of years of domination on land, these creatures suddenly started to die out all over the Earth. Some researchers believe this happened in a matter of months. Current theories suggest some kind of catastrophic climate-changing event: the crash-landing of a giant meteorite in the Yucatán Peninsula in Mexico, sending so much debris into the atmosphere that it blocked the sun's light and caused the planet's surface to cool; or the eruption of a large field of volcanoes called the Deccan Traps in central India which had the same effect through ash emissions. We will never know, but the end result was the same: from around 65 million years ago, dinosaurs disappear from the fossil record. In their place rose a diversity of birds—the dinosaurs' closest relatives on Earth—and warm-blooded mammals, which would eventually become the dominant species.

In what would become Aotearoa, the dinosaurs vanished too. And as the mountaintops dipped below and rose above the waves, unique flora and fauna began to develop. While the largest of the dinosaurs had been wiped out, the smaller ones survived, and they developed in marvellous and innovative ways. For 65 million years, Aotearoa was a land of birds.

Little bush moa (*Anomalopteryx didiformis*)

2. The biggest birds: Moa

William Colenso was a fit man, well used to walking long distances through rough country, but even he was puffing by the time he and his group of Māori guides reached the village. All work stopped as the black-hatted, pink-faced Pākehā arrived outside the wharepuni and put down the pīkau he was carrying.

Not many of the pale visitors made it this far inland, and this one looked like another mihinare. But he didn't seem interested in telling stories about *his* atua. He wanted to know *their* stories.

'Tell me about this bird,' he said, in reo Māori.

'Oh, we are not sure it *is* a bird,' replied one man. 'It has the body of a bird, but the face of a man. It is a monster.'

'It lives in a cave, up on that mountain,' joined in another man. He pointed into the distance at the peak of Whakapūnake, to the south. 'It is guarded by two tuatara, who watch over it when it sleeps. If you go there and wake it up, it will trample you to death.'

Colenso nodded, scribbling away with the stub of a pencil in a little notebook.

'And what does it eat?'

The local men looked at one another, surprised at the question.

'It doesn't eat anything—it lives off the air.'

Colenso frowned. This was sounding more and more fanciful.

'And is there more than one of these great birds?'

Again, glances were exchanged.

'Not any more.'

The missionary raised his eyebrows in query.

'There is only one that lives there on the mountain,' the first man offered. 'But we have found many bones. Enormous bones.' He spread his arms wide, as did several of the others.

'So, can you take me to see this great bird of yours?' Colenso asked. At this there was much head-shaking. They had never seen the fearful bird themselves, nor did they want to. The Pākehā was crazy to want to go anywhere near this monster.

William Colenso first visited the East Cape region of the North Island in 1838, two years before the signing of the Treaty of Waitangi made Aotearoa New Zealand a British colony. Colenso, who had come to the Bay of Islands four years earlier as a printer for the Church Missionary Society, travelled to Te Tairāwhiti with his colleagues William Williams and Richard Matthews. They were hoping to find somewhere for Christian missionaries to live and work, but Colenso was also fascinated by the people he met and the native plants and birds he encountered. He was particularly taken by the local people's stories of a giant bird which had once lived in the region.

Colenso was a typical man of his time—a period in Western history when science and reason were becoming more important than simply following the traditional teachings of the church. It was no longer enough to accept the world and all its wonders as the creations of the Christian God. European seafarers were encountering new lands and their people. Species which had never before been seen or even imagined were being revealed to the eyes of Europeans in the 'new worlds' of the Americas and the Pacific. Natural history—the study of living things—became a passion for many wealthy and educated men of this era. They wanted to discover every plant and animal—to describe, categorise and catalogue. They formed societies and published journals, describing and debating endless new discoveries.

It was also a time when the concepts of evolution and extinction were not widely known or accepted. The Earth was thought to be only thousands, not millions, of years old. But the spread of scientific thought was shaking long-held beliefs based on Christian teachings. In the 1830s and 1840s, a young geologist and botanist named Charles Darwin first formulated what would become his famous theories of evolution and natural selection. Evolution is the idea that species develop and change over time, and can even disappear entirely if they do not adapt to changing circumstances (natural selection). Darwin's theories took some time to gain acceptance after the publication of his groundbreaking book *On the Origin of Species* in 1859. (Darwin visited New Zealand during his round-the-world voyage on the *Beagle* in 1835, but found it 'not a pleasant place'.)

Although a religious man, Colenso was also a nature lover and a seeker of truth. But he wasn't the first European to hear of the large bird which once lived in his adopted country. A trader called Joel Polack, who had set up shop at Kororāreka (Russell) in the Bay of Islands, also travelled to the East Cape region in the early 1830s. In his 1838 book *New Zealand: Being a Narrative of Travels and Adventures* he recorded being told of 'a species of emu, or a bird of the genus Struthio' (that is, like an ostrich) which had once lived in the region. He was shown several large bones, and he said Māori had told him that the bird could still be found in the South Island 'in parts which, perhaps, have never yet been trodden by man'.

However, neither Polack nor Colenso got the credit for 'discovering' this strange bird. Instead, that glory went to a powerful scientist on the other side of the world.

That such gigantic birds had once stalked Aotearoa was, of course, not news to the Māori. Moa had been a critical part of Māori culture in the centuries after the first Polynesians arrived on these shores, about 800 years ago. While 'moa' was the name recorded by Colenso for the giant bird of the East Cape, the term was not widely used by Māori, and was probably spread by the missionaries themselves. 'Moa' is the Tahitian word for domestic chicken, and may have been used by some Māori to describe this much bigger bird to Europeans. One of the first written records of the bird refers to it as a 'movie', which may have been a corruption of the name Māui. Another account, written in the 1910s, suggested that the bird was once known as 'te kura', or the red bird. Other names have been lost.

Traditions and stories about hunting the bird were mostly lost, too. While moa may once have lived throughout Aotearoa, by the time Europeans started recording Māori traditional stories, these spectacular birds had been extinct for many generations. The memory of them had largely been lost.

However, archaeological evidence makes it certain that the first Polynesian arrivals made the most of the abundant resource provided by the various species of moa which roamed coastal areas, bushland and the alpine zones. For a people who had come to Aotearoa from islands rich in natural resources, as well as domesticated stock such as chickens and pigs, the hefty moa must have seemed a nutritious dietary godsend.

Moa didn't provide only meat to the first human inhabitants of Aotearoa. Their eggs were eaten, too; their feathers were used for clothing, woven into fine cloaks; and their bones were carved into tools, ornaments and fishing lures. They also seem to have served a ceremonial purpose: blown moa eggs have been found in graves at Kaikōura and the large archaeological site at Wairau Bar / Te Pokohiwi.

Bones of all species of moa, hunting and cooking tools, and the remains of moa-hunting camps have been found throughout the North and South islands. At Wairau Bar, considered to be one of the first places in Aotearoa settled by Polynesians, archaeologists have found enough bones to suggest that more than 4000 moa were killed and eaten around the area. The early Māori were so efficient at hunting moa that it is thought that all species were essentially wiped out within a century or so of the first human arrival in Aotearoa, and certainly by the late 1500s. By the time Europeans started to collect its strange, colossal bones, no one alive had ever seen one, and the giant bird had largely faded from human consciousness.

Professor Richard Owen never visited New Zealand, but his name has become forever associated with one of our most iconic birds. Owen famously publicly declared in 1839 that in New Zealand 'there has existed, if there does not now exist . . . a Struthious [ostrich-like] bird nearly, if not quite, equal in size to the Ostrich'. He was so sure, he told members of the scientific community, that he was willing to risk his reputation on it. Fortunately for Owen, it turned out that he was right.

In the late 1830s, Owen had the elaborate title of Hunterian Professor of Comparative Anatomy and Physiology at London's Royal College of Surgeons. He was also the assistant conservator of the college's museum, which had moved into large new premises in 1837.

In 1839, Owen received a letter from a Sydney doctor named John Rule, in which Rule offered to sell him 'a portion or fragment of a bone . . . part of the femur of a Bird now considered wholly extinct'. Rule had been given the bone by his nephew, John Harris, who had visited him from his whaling and trading station at Tūranganui-a-Kiwa, where Gisborne now stands. Harris told his uncle that the bone had been found buried in the banks of a river and came from a bird the local people called a 'Movie'.

Rule was at that time in London, and he took the bone to the museum to try to sell it to Owen. Owen was unimpressed. At first sight he thought it was most likely a marrowbone from a farm animal, but then he took a closer look. The bone was not solid, or smooth inside, like that of a mammal. It had a strangely textured interior, more like the bones of an ostrich. Also, it wasn't a fossil—it appeared to be a few centuries old at most. Owen compared it with other bones in the Hunterian Museum and came to the conclusion that the story Rule had told him about a great bird which had once lived in New Zealand could be true.

With Rule's permission, Owen presented the bone and his findings to the next meeting of the Zoological Society in London. While some of Owen's colleagues were dubious, he went on to publish a further paper about his 'discovery' in 1840 and put out a call for more bones to be found—or, even better, a live specimen. He wasn't prepared to rule out the possibility that these giant birds survived somewhere in the unexplored inner regions of the wild islands of Aotearoa.

It was unfortunate for Colenso that Owen was

both so well connected and physically on the spot in London, with the ear of the other notable scientists of the time. Colenso might have been on the ground in New Zealand, gathering more bones as he voyaged through the rugged interior of the North Island on foot in 1841–42, but Owen was in the spotlight.

After returning to the East Cape and collecting more strange and gigantic bones, Colenso wrote a paper for the new Tasmanian Natural History Society outlining his findings and theories. Colenso's article described a large number of bones brought by Māori to Colenso's fellow missionary William Williams, which were to be sent to London. After trying to reassemble the legs of a skeleton, Colenso and Williams estimated that the bird 'must have measured in altitude, when alive . . . from fourteen to sixteen feet [4.3–4.9 metres]'—an enormous feathered monster. The moa, Colenso surmised, was almost certainly extinct. Now the far-flung islands of New Zealand had become a British colony, he added, 'rest assured that if such an animal exists, it cannot much longer remain concealed'.

Colenso's paper included a dubious third-hand account of a recent moa sighting in Te Koko-a-Kupe / Cloudy Bay in Marlborough by two Americans, but much of its science was sound, and it included detailed drawings. It should have been a major scientific coup. Unfortunately for Colenso, the publication of his paper was delayed—the society chose to print his other submission, on ferns, first. Meanwhile, on the other side of the world, Owen had let the moa out of the bag.

In January 1843, Owen received a shipment of two large crates of moa bones from Williams. Included in the stupendous haul were several massive complete leg bones, which matched the earlier bone fragment Owen had seen and demonstrated the true size of the bird. Owen finally had the evidence he needed—thanks to Colenso and his missionary colleagues. Five days later he addressed the Zoological Society and announced the existence of a new species, which he named *Dinornis novaezealandiae*—the great, terrible, prodigious bird of New Zealand. It was the largest bird the world had ever seen.

One of the things nineteenth-century scientists studying these strange bones quickly discovered was that they varied hugely in size. While all the bones were large, not all were truly enormous. During the rush in the 1800s and early 1900s to find moa remains and classify them, up to 64 different species were identified and named, based on these differences. Over time, this number has been whittled down through closer analysis and DNA testing.

Scientists now agree that there were nine species of moa, and they had different distributions and habitats. All belong to the scientific order Dinornithiformes, but are divided into three families. In the first family are the tallest moa, the North and South Island giant moa, which were given the scientific names *Dinornis novaezealandiae* and *Dinornis robustus*. In a second family are the more stocky types: the little bush moa (*Anomalopteryx didiformis*), the eastern moa (*Emeus crassus*) and the stout-legged moa (*Euryapteryx curtus*), as well as the heavy-footed moa (*Pachyornis elephantopus*), Mantell's moa, named after Walter Mantell (*Pachyornis geranoides*), and the crested moa (*Pachyornis australis*). The third family has just one member: the slender, more athletic upland moa (*Megalapteryx didinus*).

Of these, the giant moa is perhaps the most famous. However, the birds varied greatly in size, ranging from the size of a large turkey (stout-legged moa) to around 2 metres tall in the middle of their back (giant moa). There wasn't only a size difference between species, either: female moa were often larger than males of the same species. Female South Island giant moa were the biggest of all, with their long necks reaching up to 3.6 metres high—about twice the height of an adult man! Moa had very small heads compared with their bulky bodies, and scientists believe they had a good sense of smell but poor eyesight.

Although they differed in size, shape and habitat, all the moa share one unique feature: they are the only bird ever to have existed which had absolutely no wings. Unlike other flightless birds such as the kākāpō and takahē, which have small and ineffectual wings, or even the kiwi, which has stumpy remnant wing structures tucked under its feathers, none of the species of moa had any wings at all. (Interestingly, because the moa had no breastbone, or keel, to which their wing muscles were attached, they also didn't have much meat on this part of their bodies. The most tasty and plentiful meat was in the legs, and remains of moa-butchery sites show these were the main part of the bird eaten by early Māori.)

All species of moa were vegetarian, eating the leaves and twigs of bushes and shrubs. Subfossil (partly fossilised) remains have shown that the various species lived throughout New Zealand, on the two main islands and on Great Barrier Island (Aotea) and Stewart Island / Rakiura, but not on the Chatham Islands. The largest of the birds, the *Dinornis* species, were found throughout Aotearoa, but the smaller, stockier types, such as the stout-legged moa and Mantell's moa, lived mostly in lowland and coastal areas (and Mantell's moa lived only in the North Island). The eastern moa occupied areas of lowland and swamp forest in the eastern South Island and, as its name suggests, the little bush moa was found in thicker forest. Upland and crested moa were adapted to live at higher altitudes, feeding on alpine plants on the lower slopes of the Southern Alps / Kā Tiritiri o te Moana. More than one moa species were found in some areas.

The first European scientists believed that the birds walked upright, with their heads high in the air like giraffes, and early skeletons were put together this way. However, it is now believed that moa had a humped back, more like an emu, and walked with their necks curved and extended forward, pushing through bush and undergrowth to find food. Until the arrival of humans, the only natural enemy of the South Island moa was the massive Haast's eagle. The bones of this colossal predator are sometimes found alongside moa remains; the hunter and the hunted preserved together. (For more on this flying monster, see chapter 3.)

As European settlers arrived in Aotearoa in the 1800s, felling forests, draining swamps and exploring the back country, more and more moa bones came to light all over the country. Moa, it seemed, had once been abundant. One of the most significant early finds was at Glenmark, near Waipara in North Canterbury, where in the mid-1860s workers uncovered many bones as they drained a swamp. Geologist Julius von Haast, who was collecting specimens to establish a museum for Canterbury, instructed the workers to excavate the bones as they dug the drains, and send them to Christchurch by wagon. Thousands of bones were uncovered, many of which Haast traded with other institutions around the world in exchange for specimens for his new museum.

Then, in the 1890s, a farmer discovered large bones in a sinkhole at the Kapua Swamp near Waimate in South Canterbury. Canterbury Museum paid just £40 for the right to dig up the swamp, and workers proceeded to uncover a massive haul of lost moa. Photographs of the dig show hundreds of bones stacked like firewood, ready to be loaded on to railway wagons and carted off to Christchurch. It is estimated that more than 2 million bones were dug from the swamp, although many were broken or damaged and only the best specimens were retained.

Two of the most famous moa discoveries came in the late 1930s. In 1937 at Pyramid Valley, again near Waipara, two farmers were digging a hole in a swamp to bury a dead carthorse when they came across three massive moa leg bones. They put them in the woolshed, thinking that the Canterbury Museum already had enough moa bones in its collection. The next year, a visiting naturalist saw the bones and realised that they came from a single moa, unlike the jumbled masses of mixed-up skeletons which had often been found elsewhere. He got the museum interested, and a 'moa rush' ensued.

Excavations of the swamp during and after World War II uncovered a large number of skeletons, many complete, of four different types of moa, as well as the remains of other extinct birds such as the native goose and Haast's eagle. It appeared that over a long period of time, hundreds of unfortunate moa had wandered on to an area of peat swamp with a thin crust, then sank down into it. They had become trapped and they died where they stood, unable to clamber out without wings. But bad news for the moa was good news for scientists: following the Pyramid Valley finds, they were better able to classify the different moa species.

Of similar significance was a discovery made by a schoolboy, Jim Eyles, in 1939. Eyles had been brought up on Boulder Bank, the long strip of sandy, gravelly soil that forms the Wairau Bar / Te Pokohiwi, where the Wairau River meets the sea on the coast of Marlborough. Before Eyles was born, his father had uncovered bones and other artefacts when ploughing a paddock to plant peas, and Eyles had grown up with a passion for 'fossicking'—searching for buried treasures.

One day in January 1939, just after his thirteenth birthday, Eyles and a mate were poking about looking for 'curios'—rare and unusual objects—just 100 metres from his house when Eyles found a moa egg, virtually intact except for where he'd accidentally sliced the end off it with his spade. A bit more digging revealed a collection of human bones, including a skull and a beautiful and unusual necklace, made of reels of whale ivory. The finds were displayed in the window of the local fish shop in Blenheim until they were bought by the Dominion Museum (now Te Papa).

No further digging was done at Wairau Bar until 1942, when Eyles found several more burial sites containing strange and wonderful treasures. Roger Duff, an ethnologist at the Canterbury Museum, became interested, and together they began to excavate what became New Zealand's most significant archaeological site. It was not so much the moa bones found at Te Pokohiwi that were of such interest: it was the evidence of very early human habitation, which provided a previously missing link between the first Polynesian arrivals and the recognisably Māori culture that developed. In 2009, the kōiwi tangata (human remains) and many of the grave goods found in the 1940s were returned to the site and reburied.

Today, moa bones and other artefacts continue to come to light. In 2001, contractors digging a drain at the Bell Hill vineyard, not far from the Glenmark and Pyramid Valley excavations at Waipara, uncovered a large swamp site containing the bones of many extinct birds, which is said to be one of the most significant finds of the past century. And in 2016, electricity contractors digging a trench in South Canterbury uncovered the remains of a moa 'family'—an adult male and female and a younger male bird.

While a lot of what scientists know today about

moa comes from their bones, many other mummified and subfossil remains have also been found. Huge eggs, moa feathers, skin and muscles, and even whole heads and feet have been discovered, dried out in caves. Archaeologists and fossickers have also found many gizzard stones—rocks swallowed by moa and kept in part of their stomach to help them grind up and digest tough plant material—and even fossilised moa poo. Moa left other traces, too: in early 2019, Otago tractor driver Michael Johnston discovered the first preserved moa footprints ever found in the South Island, in the bed of the Kyeburn River in the Maniototo.

Most scientists now believe that the last moa disappeared in the 1400s, although the crested moa may have held out for longer in the isolated upland of northwest Nelson (Whakatū). Although moa have been extinct for hundreds of years, some people believe that the birds still exist, in remnant populations—small groups of survivors—hidden away in isolated areas such as Te Urewera, the Southern Alps / Kā Tiritiri o te Moana and Fiordland (Te Whakataka Kārehu a Tamatea). Every few decades a sighting is reported, and cryptozoologists (people who search for and study animals which are not proven to exist) mount expeditions in search of further evidence.

While many sightings over the years since European settlement can be dismissed as deliberate hoaxes or accidental flights of fancy, several have a ring of truth about them. A young girl living in remote South Westland in the 1880s, Alice McKenzie, wrote in later life of seeing a large bird on the beach near her home, and finding unusual footprints several times. While some believe this to have been a takahē, the size of its tracks and the colour of its legs (grey or green, not red) suggest that it could have been a moa. As recently as 1993, international media interest was piqued by Arthur's Pass publican Paddy Freaney. Freaney, along with two tramping companions, claimed to have seen a large grey and brown bird standing about a metre high, near a river in the Craigieburn Range. Freaney was able to take a blurry photograph of the animal as it fled. Some experts believed it showed a red deer, but Freaney was adamant it was a moa, and he defended his story for many years until his death in 2012.

While a live moa will probably never be seen again by human eyes, there is a possibility that one day the birds will be resurrected by science. In 2018, researchers at Harvard University in the United States announced that they had put together a nearly complete genome (set of chromosomes) of the little bush moa after extracting its ancient DNA from a single preserved toe bone. While we are a long way from seeing newborn cloned moa breaking out of a giant egg, the 'de-extinction' of moa and other lost species is no longer solely in the realm of science fiction.

In the meantime, we may have to be satisfied with the words of this folk song. For now, it seems there are:

No moa, no moa,
in old Aotearoa
Can't get 'em,
they et 'em
They're gone and there ain't no moa.

Haast's eagle (*Aquila moorei*) attacking a moa

3. Terror in the skies: Haast's eagle

It had not been a good day's hunting. The two men had walked all day and found few signs of any moa. There had been a pair of large geese down by the lake edge, but the sound of the hunters' approach had scared the birds away. The men were hungry and tired, and they knew that when they returned to their village the chief would not be pleased with them.

Suddenly, one of the men stopped in his tracks. A strange noise was coming through the bush. It definitely wasn't human, but it also didn't sound like any bird call they had ever heard. It was a loud screech—a high-pitched cry. The bird sounded as though it was scared or in pain.

As quietly as they could, the two hunters pushed through the underbrush in the direction of the sound. As they reached the edge of the trees, the men could see that the noise was being made by a moa, standing about halfway across an area of low-lying swampy ground. The moa's scaly feet had sunk into the soft earth and it was struggling to free itself, thrashing about with its neck as it tried to tug its legs free. But the more the bird panicked, the further it sank into the mire.

The hunters grinned at each other. This would be an easy kill. They would just need to be careful not to get bogged down themselves.

But as they raised their spears and stepped out from the edge of the forest, a shadow fell over them and they recoiled in terror. A gigantic bird, its wingspan twice as wide as a man was tall, swooped out of the sky. Its talons were raised, ready to attack, and a harsh cry came from its viciously hooked beak.

'Pouākai!'

Both men dived for cover, but they were not the eagle's intended prey. Instead, the huge bird dropped down on to the back of the trapped moa and sank its claws into the bird's feathered sides. The moa screamed in pain, swinging its long neck around and trying to bite its attacker. But it was no match for the bird of prey. The eagle grabbed the moa by the head with its beak, poking it through the moa's eye sockets and into its brain. The moa struggled for a little longer, then its hefty body flopped to the ground. The eagle then started to calmly tear great chunks of flesh from the still-warm body, never once looking around at the two men cowering in the bushes nearby.

It had been a bad day's hunting all right, but not for the eagle.

The first Polynesian people to settle in Aotearoa New Zealand found a country unlike what they had been used to in the tropical Pacific islands. It was bush-clad and mountainous, with more extreme seasons. And the first Polynesian voyagers who came to these islands found not only birds to eat, but also birds that might eat them.

Māori legends tell of a huge bird that once lived in the mountains, which would swoop down to carry off the unwary. In 1872, Sir George Grey, British governor of the new colony of New Zealand from 1845–54 and 1861–68, recorded a story he had been told by Māori of an 'extinct giant bird of prey'—a bird that hunts animals—called hōkioi. This bird, Grey had been told, was a very large, powerful hawk that was rarely seen. It was said to have black feathers 'tinged with yellow and green [and] a bunch of red feathers on the top of its head'. A Māori legend reported by Grey to the Wellington Philosophical Society told of its rivalry with the hawk, kāhu:

> The hawk said that it could reach the heavens; the hokioi said it could reach the heavens; there was a contention between them. The hokioi said to the hawk, 'what shall be your sign?' The hawk replied, 'kei' (the peculiar cry of the hawk). Then the hawk asked, 'what is to be your sign?' The hokioi replied, 'hokioi-hokioi-hu-u.' . . . They flew and approached the heavens. The winds and the clouds came. The hawk called out 'kei' and descended, it could go no further on account of the winds and the clouds, but the hokioi disappeared into the heavens.

Legends from other parts of Aotearoa told of a huge bird called pouākai. Missionary James Stack was told by South Island Māori of the traditions of the Waitaha people, the first inhabitants of the island, about pouākai. One of these huge birds had built its nest on a mountain in the foothills of the Southern Alps / Kā Tiritiri o te Moana and would dart down to seize and carry off men, women and children as food for itself and its young. Although its wings made a loud noise in flight, 'it rushed with such rapidity upon its prey that none could escape from its talons'. The name pouākai meant 'old glutton', referring to the bird's insatiable appetite.

Stack wrote that a visiting warrior eventually volunteered to try to rid the local people of this pestilence. He made a trap by covering a pit with a network of mānuka saplings and hid within it a force of 50 men armed with spears and 'thrusting weapons'.

> [H]e himself as soon as it was light, went out to lure the Pouakai from its nest. He did not go far before that destroyer spied him, and swooped down upon him. Hautere [the warrior] had now to run for his life, and just succeeded in reaching the shelter of the network when the bird pounced upon him, and in its violent efforts to reach its prey, forced its legs through the meshes, and becoming entangled, the fifty men plunged their spears into its body and after a desperate encounter succeeded in killing it.

As well as these oral histories there is visual evidence, too. Several paintings in caves and rock shelters in the South Island appear to show giant eagles, alongside depictions of people and moa. Ornaments and tools fashioned from eagle bones have also been found in Te Ana o Hineraki (Moa Bone Point Cave) at Sumner, Christchurch, and elsewhere eagle bones have been found in old middens (piles of food rubbish)—suggesting that, sometimes, the predator may have become the prey.

The first solid evidence that such a creature had indeed once terrorised the skies of Aotearoa was found in 1871. Perhaps not surprisingly, the first eagle bones found by Europeans were uncovered alongside the remains of its primary prey, the moa. Workers excavating a swamp filled with moa bones at Glenmark, near Waipara in North Canterbury, in the 1860s uncovered many smaller bones along with the big ones. Julius von Haast, the director of the new Canterbury Museum, sent the museum's taxidermist, Frederick Fuller, to keep an eye on the excavation. It was Fuller who first realised in 1871 that among the scattered heaps of moa bones were the remains of other birds—including two huge and terrifying claws which could only have belonged to a bird of prey.

Fuller trained as a taxidermist in England before emigrating, first to Australia in 1849, then on to Otago with his new wife in 1862. There, he started working with Haast, collecting and stuffing specimens of South Island birds and fulfilling his boss's goal of having a 'complete set' of all the birds living in the Canterbury province. Fuller was working for Haast when the swamp was discovered at Glenmark, and together the two men pieced together the puzzle of moa bones to create a series of skeletons to go on display at the new museum, which had opened to much wonderment in 1867.

Haast was excited by Fuller's discovery of the eagle bones. He took the unusual step of identifying and naming the species himself rather than sending the bones to England to be categorised (perhaps he had bad memories of the specimens which had been lost when the ship *Matoaka* sank a few years earlier—see chapter 1). Haast proposed to give the gigantic bird the scientific name *Harpagornis moorei* (now *Aquila moorei*), from the Greek word meaning 'seizer or robber' (harpax) and the name of the owner of the Glenmark property, George Henry Moore. In his first scientific paper on the new eagle discovery, however, Haast gave Fuller the credit he was due for the find.

Haast surmised that the eagle was active during the day, like the harrier hawk or kāhu (*Circus approximans*), and 'doubtless followed the flocks of Moas, feeding either upon the carcases of the dead birds, or killing the young and disabled ones'. Later excavations found evidence of claw marks in the pelvises of full-grown moa, suggesting that the eagle took on even the largest of these birds, sinking its claws into their haunches and tucking in to their kidneys.

Further adding to the sense of the size and fearsomeness of this new discovery, Haast added:

> Only the lion and tiger, amongst the recent carnivorous mammalia, perhaps have larger ungual phalanges [toe bones ending in a claw] than this extinct raptorial bird, and after having seen its curved talons, the fable of the bird Roc

[a giant bird of Middle Eastern mythology] no longer seems so very extravagant and strange, and I may add, that a human being, if not well armed or very powerful, not to speak of children, would have stood a very poor chance against such a formidable foe, if it had chosen to attack him.

Two years after his first scientific paper on the eagle was delivered in 1871, Haast provided the members of the Philosophical Institute of Canterbury with an update. More bones had been excavated at Glenmark, and a well-preserved eagle pelvis had been found in a rock shelter in Central Otago by a gold prospector. This mummified remnant still had some muscles and ligaments attached, although the bird it belonged to had died many hundreds of years previously.

While Fuller was credited with the initial discovery, it was Haast's name which became attached to the eagle, and Fuller's contribution largely disappeared from view. Fired from the museum due to his drinking problem, Fuller took some of the arsenic he used to preserve bird skins and died of poisoning in July 1876.

We now know, from further fossil bone finds, including at Pyramid Valley in North Canterbury (see chapter 2 for more on that excavation), that New Zealand's extinct giant eagle weighed up to around 18 kilograms and had a wingspan of around 3 metres—not much smaller than a Mini car is long. It had hugely powerful legs and sharp claws, which it used to kill and dismember its prey. Its hooked beak, like that of a vulture, led scientists to contend for some years that it was a scavenger, feeding on dead birds, but they now believe it was a predator, killing quite large birds such as moa and using its beak to tear them apart to eat them.

Recent DNA analysis has suggested that the giant eagle's closest relative is in fact the Australian little eagle, which weighs only around 1 kilogram. They seem to share a common ancestor just a few million years ago—very recent in evolutionary terms. This research suggests that the giant eagle's ancestors came from across the Tasman and became very large in a very short period of time, in an environment rich in prey and devoid of predators.

Studies show that the eagles were probably never abundant when humans first lived in the South Island. One population estimate, based on fossil remains, suggests that there were only around 3000–4500 breeding pairs in prehistoric times. There are Māori stories of the pouākai from the North Island, but so far eagle bones have been found only in the South Island and Stewart Island / Rakiura. (The North Island had its own top-of-the-food-chain predator: Eyles' or Forbes' harrier, *Circus teauteensis*, which also became extinct after the arrival of Polynesians.)

So what became of this giant eagle, a terrifying presence which must have feared nothing? It seems that when the first humans arrived in its territory after thousands of years, the eagle finally met its match. While the first Polynesian settlers probably didn't kill many of the huge birds directly, their actions led to its demise. As the forest cover was burned off the Canterbury Plains, the eagle's habitat shrank. The newcomers also competed with the predator for its favoured food—the moa. Within a few hundred years of human settlement, both the eagle and its prey had vanished.

The eagle and Eyles' harrier were not the only casualties of human hunting in these first few hundred years. Aotearoa was also once home to a large flightless goose that stood about 70 centimetres tall; several different types of duck, some of which may also have been unable to fly; another waterbird called a coot; a raven that foraged around the coast; and a large flightless bird called the adzebill that, as its name suggests, had a large, downward-curving beak. It may have preyed on lizards including tuatara and small birds. All these birds, many of which had lost the ability to fly and had no fear of humans, would have been easy pickings for Polynesian hunters, and the large size of the geese and adzebills would have made them a generous meal. The smaller coot seems to have been one of the most popular birds targeted by hunters; excavations have uncovered hundreds of coot bones in midden sites throughout the country.

Other smaller species fell prey to the kiore, the Pacific rats which travelled along with the Polynesians in their waka. The kiore (*Rattus exulans*) was smaller than the European rats which would soon invade its territory, and not such a good swimmer or climber. While its diet was mostly seeds, berries and fruits, it also ate insects, lizards, bird eggs and chicks, and is believed to have been responsible for the extinction of several smaller species, such as the flightless stout-legged and long-billed wrens. These tiny birds simply had no defences against such a predator.

Some birds survived the human and animal hunting and the changes to their habitats, however, and when the first European ships started to call in Aotearoa, the forests were still full of birdsong. It wasn't long, however, before more species were silenced forever.

South Island piopio (*Turnagra capensis*)

4. New killers take their toll: Whēkau, koreke, piopio

Canterbury Settlement
May 4, 1854

My dearest Mother,
We have now been a month in the new settlement and slowly I am growing accustomed to our surroundings. Canterbury is a very flat place, with grassy plains as far as the eye can see, all the way to the mountains in the distance.

There are also very big hills which separate the settlement from where the ships arrive, at Port Cooper. When we first came ashore we had to walk over them, while much of our luggage went around the sea way into the river which runs up to the settlement. So far there are not many houses built, but there are a few of a passable quality. There is very little timber here, the forests on these plains having been burned off long ago by the Maori people.

Though we are in a house now, when we first arrived we were living in what they call a V-hut, which looks exactly like that—an upside-down V, with a roof of thatch stretching right down to the ground on each side, a door and a window at one end, and room to stand up only at the highest point. All our cooking was done outside, and that was a rough job in the weather we have here, I can tell you.

Altogether it is very strange. The country is brown, not green, and it is very rough beyond where the settlers have tried to make their gardens. There are no flowers—none of the native bushes seem to produce anything pretty, and around where we are living it is mostly tussock grass.

There are so few birds here, compared with Home: no blackbird or even a sparrow. There is a native thrush, that they call piopio, but it lives only in the bush, not in our gardens. At night we sometimes hear owls, but they have a strange call, like a wicked laugh. There are also quail here, a bit bigger than what we are used to. They are good eating and make a fine dinner—a pleasant change from the mutton which is our usual fare. George bagged half a dozen quail last Wednesday, though he says these birds are already becoming more scarce and harder to shoot . . .

Extinction was a concept unknown to the first European settlers who came to the colony of New Zealand in the mid-1800s. Most of them came from countries which had been settled by humans for thousands of years, where the animals, birds and plants had adapted to sharing their environment. Predators and prey lived alongside each other, with populations in balance.

The idea of a whole species or type of animal dying out and disappearing from the world due to the actions of humans was new, and it was just starting to occur in lands being colonised by Europeans. In Mauritius in the Indian Ocean, the dodo was lost for all time in the 1600s, following the arrival of Dutch settlers. Over the Atlantic Ocean in the Americas, creatures new to Western eyes were decimated by the spread of European settlement in the 1700s and 1800s, and species such as the bison were hunted nearly to extinction. A unique type of marine mammal named Steller's sea cow, discovered by shipwrecked Russian sailors in the Bering Strait between Russia and Alaska in 1741, was polished off by hungry fur traders and seal hunters within 30 years of its discovery by Europeans. In each case, animals or birds adapted to their own ecosystem were no match for the destructive tendencies of the most powerful predator on Earth: humans.

The same scenario played out in the far-flung islands of Aotearoa. The new settlers at times appreciated and at worst simply tolerated the multitude of beautiful and unusual bird species with which they found themselves sharing their new home. They may have written letters home to their families like the imagined one at the start of this chapter, noting the strange new birds they encountered here. However, they were largely unaware or unthinking of the effect that their presence had on the original occupants of the land. Extinction of native species, which had slowed from the early centuries of Polynesian settlement, now ramped up again in the first fifty years of European settlement of the islands of Aotearoa.

When the Treaty of Waitangi was signed in February 1840, Aotearoa was sparsely populated. Estimates suggest that around 100,000 Māori lived throughout the three main islands, Te Ika-a-Māui, Te Waipounamu and Rakiura, at the time of Captain

James Cook's first visit in 1769. By 1840 the Māori population was around 80,000, but due to the spread of new diseases it continued to fall through the 1840s and 1850s. By the 1870s there were around 300,000 European New Zealanders. They spread out from port settlements such as Auckland, Wellington, Christchurch and Dunedin into a previously largely untouched hinterland.

And it wasn't just the people who put pressure on the pristine natural environment that they encountered, as they felled and burned trees to create farmland—it was the animals and plants they brought with them. The first animal arrivals to Aotearoa had come with the Polynesian voyagers—rats (kiore) and dogs (kurī). Kiore were responsible for the loss of several smaller bird species in the pre-European period, but Norway rats, which probably came with Cook and later explorers, were larger and more aggressive. They made quick work of many of New Zealand's ground-dwelling, flightless or low-nesting birds and their eggs, as well as insects and other invertebrates. They were well established by the early 1800s, and then in the 1860s they were joined by ship rats, which could swarm to plague proportions. The first European settlers also brought mice, which competed with native birds for food, and possums were introduced in the 1850s in the hope of establishing a fur industry.

In addition, Cook brought pigs and goats on his

first visit to New Zealand. He also brought shipboard cats, which wreaked havoc among the birds on shore—wild cats were already established and doing their wicked work by the 1830s. Rabbits were brought to New Zealand from the 1830s as a source of food. They did not prey on native species, but the animals released in the 1870s in an attempt to control them—more cats, stoats, weasels and ferrets—certainly did. Aotearoa's treasure box of native species was under attack on all fronts—from habitat destruction, human hunting and introduced predators. Many would not survive.

The majority of the first European settlers who came to New Zealand from the 1840s onward came from the British Isles. These people would have been familiar with owls, and perhaps they were comforted to find that their newly adopted homeland was also populated by these nocturnal birds. Today, most New Zealanders know of the ruru, or morepork, whose distinctive swooping two-note call can still be heard throughout the country. But we once had a second native owl, known as whēkau, hākoke or kakaha, the laughing owl (*Sceloglaux albifacies*).

The whēkau was about twice the size of a ruru, standing about 45 centimetres high, and had the same colouring: dark brown with cream-coloured patterning. Stuffed specimens show its short tail and large claws, with circles of paler feathers around its large eyes. In one surviving black-and-white photograph, a young bird glares balefully out of a crevice in a limestone boulder, a dead mouse hanging limply from its beak. The whēkau lived in crevices and holes in cliffs, especially in limestone country. The Māori expression 'me te pari hākoke' (like a hākoke cliff) was used to describe areas of land to be cultivated which were steep and stony. The whēkau was once found throughout the country, although by European times it was found only in the lower North Island and on the eastern side of the South Island, from Nelson down to Otago. While nineteenth-century observers describe it living mostly in open country, fossil evidence suggests that the whēkau was once a bush bird, too.

To Māori, owls were considered omens, their appearance or call thought to be a warning of coming ill fortune. The name whēkau means 'guts' or 'entrails' and refers to the bird's preferred food. Its cries could be heard on drizzly nights or before rain, so it was also nicknamed 'the rainbird'. Māori used the bird's skins as ornamental pouches and its feathers for making cloaks.

This owl's European name, the laughing owl, is a bit misleading. Contemporary reports suggest that its call was more of a yelping or cackling than a laugh. As Canterbury naturalist Thomas Potts noted, 'If its cry resembles laughter at all, it is the uncontrollable outburst, the convulsive shout of insanity.' Ornithologist and collector Walter Buller described the captive laughing owls he kept as 'habitually noisy . . . and when several of them are hunting together they seem to laugh in unison'. Buller's son described the captive birds making whistling noises, 'and sometimes a note very much like a Turkey chuckling. Another sound they produce is exactly like the mewing of a cat.' Another nineteenth-century observer suggested that the bird sometimes made a call 'something like that of the Morepork, but just as if he had his mouth full'. As well as keeping live owls himself, Buller also shipped pairs to his best customer, a wealthy British banker who dabbled in natural history named Walter Rothschild, who kept a private zoo and museum at his estate northwest of London.

Buller's owl supplier was William Smith, a gardener on a large property near Timaru who made a speciality of catching and keeping them. The birds seemed to survive well in captivity, even on long sea voyages, and would breed in captivity, too. However, a pair which Buller was hoping to send to the Zoological Society in London in the 1880s escaped when Buller was moving house.

> Through an unfortunate accident to the temporary cage in which these birds were being removed to my new residence on Wellington

> Terrace, they both escaped one stormy night, and were never seen again . . . Many persons declared having heard them, from time to time, on the neighbouring hills, and guided by these reports the fugitives were traced through Sir James Prendergast's grounds to the Episcopalian Cemetery, where the scent was hopelessly lost, although the old sexton solemnly averred he had heard 'most all kinds o' noises among them graves'.

In 1893, after returning from a trip to England, Buller told the assembled members of the Wellington Philosophical Society, 'This owl is now on the verge of extinction'. In the previous three years he had only managed to obtain a single live pair—which he had sent to Rothschild in England. However, as numbers of the birds dwindled, no one made any effort to preserve or protect them. The last official specimen was collected in 1914; a dead bird found at Blue Cliffs, southwest of Timaru.

It is now thought that predation by introduced mammals hastened the end of the whēkau. It had adapted to the presence of Polynesian rats in its environment by eating them, but the larger, more aggressive introduced rats, cats, stoats and weasels were too much for this placid, ground-feeding bird to cope with.

Isolated reports of the owl continued to be made

into the early twentieth century, but it now seems certain that the whēkau has laughed for the last time.

Another species of bird that looked familiar to homesick European settlers was the native quail, abundant flocks of which darted about the open country near the new settlements in the North and South islands. Quail were considered a game bird in Britain, and the little koreke, as they were known to Māori, became fair game here, too. The koreke was one of the first species to disappear into the realms of extinction after the establishment of European settlement in the 1840s.

The native quail was slightly smaller than the crested California quail which has now taken its place in the ecosystem—about 18–20 centimetres long, and weighing around 200 grams. The males were reddish brown with mottled dark-brown markings, and they had a chestnut-coloured face and throat. Females looked similar, but without the reddish face. Koreke were prized by Māori as a food source, and were given the name manu oho or 'treasure bird'. The quail was also considered a fortune-telling bird—a manu tohu or 'bird of signs'—and where its cry was heard from meant either good luck (the right side) or bad luck (the left). It was said to have been brought to these islands on the Aotea canoe.

Although koreke could fly, they lived mostly on

the ground. Māori caught them with long-handled nets and in snares stretched across the trails the birds made through long grass. Māori pepeha (sayings) refer to quails springing up from the grass like warriors jumping out of hiding. Koreke were considered a significant resource, and quail preserves were sometimes mentioned in Māori Land Court dealings in the nineteenth century.

Despite hunting by Māori and predation by kiore, the little quail were still widespread when the first Europeans started to arrive in Aotearoa. They were not collected and scientifically described, by the naturalists on Captain Cook's first two expeditions, in 1769–70 and 1773–74, and by Marc-Joseph Marion du Fresne's men in the Bay of Islands in 1772. Then, in 1827, French naturalists Joseph Paul Gaimard and Jean René Constant Quoy, travelling on Jules Dumont D'Urville's ship *L'Astrolabe,* killed some specimens on the Hauraki Plains near the Waihou River and took them back to Paris, France. There, the species was scientifically described and given its Latin name, *Coturnix novaezelandiae* (simply meaning New Zealand quail).

It seems that koreke were already declining before European settlement was established in New Zealand. Even in the 1840s the birds were not common in the North Island, and by the 1860s they were rarely seen there. The koreke's last stronghold appears to have been the grasslands around the Canterbury and

Nelson settlements, where the 'excessively abundant' birds were shot with impunity by the new settlers. Buller, writing in 1888 when the bird had not been seen for more than a decade, recorded that:

> The first settlers, who carried with them from the old country their traditional love of sport, enjoyed some excellent Quail-shooting for several years; and it is matter of local history that Sir D. [David] Monro and Major [Mathew] Richmond, in 1848, shot as many as forty-three brace [86 birds] in the course of a single day within a few miles of what is now the city of Nelson: while a Canterbury writer has recorded that 'in the early days, on the plains near Selwyn, a bag of twenty brace of Quail [40 birds] was not looked upon as extraordinary sport for a day's shooting.'

Buller also noted that some land-owners had attempted to 'preserve' their game resource by restricting when and where the birds could be shot. But their efforts were in vain, and the quail seemed to be vanishing. Buller was not sentimental, however; after noting the probable extinction of the quail, he added, 'Its place, however, has been more than adequately supplied by several introduced species, all of which appear to thrive well and multiply in their new home', including English and Chinese pheasants, partridge, and California and Australian quails.

It is not clear why the little koreke went extinct so rapidly. It had survived hundreds of years of predation by humans and kiore, yet the arrival of Europeans with their guns, and the feral cats and Norway rats they brought with them, spelled its sudden demise. It was gone almost before anyone noticed.

Twice since then, hope has flared that a remnant population of native quail may persist on an offshore island. In 1887, Auckland Institute and Museum curator Thomas Cheeseman became the first botanist to visit Manawatāwhi / Three Kings Islands, off the tip of the North Island, and he recorded koreke among the species he saw there.

> Its occurrence on the Three Kings was quite unexpected; and as it is nearly, if not altogether, extinct on the mainland, we may congratulate ourselves that it has been found in a locality where it is likely to remain undisturbed for many years to come.

However, Cheeseman's excitement proved to be misplaced. As soon as Buller saw one of the quail eggs that Cheeseman had brought back with him, he was able to identify the bird as an Australian brown quail.

Coincidentally, it was while on this same trip that Cheeseman heard a story about the only other bird related to the quail to have lived in New Zealand or its offshore islands. When visiting the Kermadec Islands, a volcanic archipelago lying 800–1000 kilometres northeast of New Zealand, about halfway to Tonga, Cheeseman spoke to Chris Johnson, who had lived on the largest island, Raoul Island, in the 1870s. Johnson told him of a bird that had once inhabited the floor of the island's large crater,

> which made mounds of sand and decayed leaves 2ft. or 3ft. [61–90 cm] high, laying its eggs in the mounds. He was in the habit of visiting the mounds for the sake of the eggs and young birds, and has frequently taken five or six of the latter from the same nest at one time.

Cheeseman surmised from its behaviour that the bird must have been a type of megapode, a ground-dwelling bird found in other parts of the Pacific that builds mounds to lay its eggs in. Cheeseman noted that a similar bird lived in the crater of the Tongan island of Niuafoʻou.

However, when Raoul experienced one of its many volcanic eruptions in the 1870s, the floor of the crater was covered with 'a deposit of mud very similar to that thrown out by the eruption of Tarawera [in 1886], which apparently killed out the species, for it has not been seen since.'

While Cheeseman and Johnson blamed the eruption for the loss of the megapode, many other bird species survived it and following outbreaks of ash and lava. It seems more likely that the unusual bird, like its counterparts on other Pacific islands, was killed by introduced predators such as cats and pigs, and overhunted by the whalers and other seafarers who visited the island. No physical trace of the Raoul Island megapode has ever been found.

As recently as the 2000s, the ghost of the koreke rose again, probably for the final time. PhD student Mark Seabrook-Davison studied quail living on island sanctuary Tiritiri Matangi, near Auckland, in the hope that they might be an isolated population of koreke. Seabrook-Davison analysed the DNA of quail from the island and compared it with the DNA of Australian brown quail and koreke; unfortunately, it showed the birds to be brown quail. The koreke, it seems, is gone forever.

Another avian wonder lost in the years following European settlement was a bird known to the new arrivals as the native thrush—although it was not related in the slightest to the European songbird. Māori knew the bird as piopio, and it was found widely in bush country throughout the North and South islands when the first European towns started springing up around the coast from the 1840s onwards.

The piopio is said to have come to Aotearoa on the Mātaatua canoe. The bird is also said to have accompanied the demigod Māui on the last of his legendary adventures, his quest to conquer death by entering the body of the goddess Hine-nui-te-pō. The piopio sang to Māui to keep up his courage, but it did not complete its song before Māui was crushed and killed by the goddess. For this reason the song of the piopio was considered to be only ever half-sung.

As with many other birds, the song of the piopio was considered an omen. The phrase 'E tangi mai a te manu nei a te piopio' means 'the piopio cries out', and its call was said to determine whether a person would die a natural death or in battle. Sadly, death in battle was the fate of the bird itself—a battle against introduced predators would silence its song entirely less than 100 years after European settlement.

There were two species of piopio, one which was found in the North Island and one in the South. Both have misleading Latin names: the South Island bird, collected and named by Anders Sparrman, a naturalist on Cook's second voyage to the Pacific, was given the name *Turnagra capensis* because it was mistakenly labelled as coming from the Cape of Good Hope in South Africa. The North Island bird has the doubly confusing name of *Turnagra tanagra* because it was initially considered to be part of the tanager family, a group of birds found mainly in South America. However, DNA studies have recently shown the piopio was closely related to a group of birds found in Australia.

Piopio were considered common in both Māori and early European times. The two variants were slightly larger and much bulkier than the European song thrush now common in our gardens. The South Island bird bore a passing resemblance to a thrush, with a speckled chest and brown back, wings and tail, but the North Island bird lacked speckles and was more brown and grey. Like the European thrush, the native birds were omnivorous, and fed on berries and seeds as well as insects and invertebrates such as worms.

Piopio also shared its singing voice with the European thrush. The piopio's song was described as a series of five or six distinct patterns, often repeated but interrupted by the sharp whistling cry from which it got its Māori name. Buller described it as 'unquestionably the best of our native songsters'. However, bushman Charles Edward Douglas, popularly known as Mr Explorer Douglas for his adventures around South Westland and Otago in the late 1880s, was less impressed, writing, 'It has no song and its cry is only a harsh twitter, a note the bird is obviously ashamed of, as it seldom uses it.' Douglas considered the birds to be 'a clumsey [*sic*], stupid sort of being, nothing . . . interesting in any way, except perhaps in their tameness'. As with many of Aotearoa's lost wonders, it was this tameness and lack of fear of humans or predators that would be the piopio's downfall.

In the 1870s Buller wrote that the North Island piopio was 'comparatively common in all suitable localities throughout the southern portion of the North Island, but . . . extremely rare in the country north of Waikato'. By the time he published the second edition of his *A History of the Birds of New Zealand* in 1888, piopio were already in decline. Buller described the North Island variety as 'certainly doomed to extinction within a very few years'. Of the South Island species he wrote:

> Formerly it was excessively abundant in all the elevated wooded country; but of late years it has become comparatively scarce, while in some districts it has disappeared altogether. This result is attributable, in a great measure, to the ravages of cats and dogs, to which this species, from its ground-feeding habits, falls an easy prey.

Piopio were one of the species which early conservationists spoke about saving—although their actions didn't match their words. In 1885, amateur naturalist Hugh Martin proposed to the Nelson Philosophical Society that a range of native birds, including both varieties of piopio, be urgently preserved in reserves. (He also mentioned the koreke and whēkau, although one was probably already extinct and the other nearly so.) Martin was one of the first to suggest that islands would be the best places of refuge, so that birds 'should be as much as is possible protected against the ravages of beasts of prey'. He wrote:

> I would again call attention to the necessity of immediate action, on account of the opening up of the back country, the rapid increase of population, and last, but by no means least, the introduction of the weazel [*sic*] and other vermin, which must on the mainland certainly lead to the destruction of all ground-birds.

In the early 1890s, the government made moves to establish Te Hauturu-o-Toi / Little Barrier Island in the Hauraki Gulf and Resolution Island in Fiordland as nature reserves. In 1893 Buller wrote:

> It is to be hoped that when the Little Barrier Island has been acquired by the Government for the purposes of a 'native birds' sanctuary' at least one pair of the North Island Thrush (which is easily caught when found) may be obtained, and liberated there, so as to save the species from ultimate extinction.

One of the final strongholds of the South Island piopio was Stephens Island (Takapourewa), a rocky island off the northern tip of Rangitoto ki te Tonga / D'Urville Island in the western Marlborough Sounds.

The birds were still plentiful in the early 1890s when workers began constructing a lighthouse there—there may have been as many as 1000 birds on the small island—but they were wiped out by feral cats before the end of the century (for more on this sad story, see chapter 6). Like Martin, Buller noted that 'small island sanctuaries' might be the only way to save the birds from extinction—failing to note that they had, in fact, been living on one until the arrival of the cats.

However, the end was near. In 1905 Buller wrote that the South Island piopio was 'fast expiring'; and that was the last year one was seen. In 1902, a North Island piopio was shot at Ōhura in the King Country, not far south of the small town which still bears its name. It seems that the piopio's song will remain half-finished forever.

Huia (*Heteralocha acutirostris*)

5. The dying song: Huia

Walter Buller crouched in the ferns. His back hurt and his knees were creaking, but he had to hold himself as still as possible. For the last hour he had heard the cries of the huia in the forest around him, always just out of sight, but now one of the magnificent birds had fluttered out of the bush and landed on a rotting log just 30 feet away. He felt a rush of excitement flood through his cramped muscles.

The bird's short, pointed, ivory-coloured beak indicated that it was a male. It twittered softly as it sprang about on the rough surface of the log, making darting stabs at the wood. It tilted its head to peer inside the cracks and let out a soft whistle. Buller knew it was calling to its mate. It was only a matter of time.

Sure enough, a second glossy black bird with a pair of orange wattles at her cheeks bounded into view, dropping down silently from the tree canopy to land next to her mate. The birds were roughly the same size, but there was one stunning difference: the female's beak was much longer than her mate's, and gracefully curved. She joined him in probing into the soft surface of the rotten bark. The pair hopped back and forth, tugging away pieces of timber to reveal fat, creamy grubs hiding within. As they moved around, they used their long tails, tipped with white, to keep their balance.

The birds were beautiful—in prime condition, with shiny black eyes and immaculate feathers. Buller knew what a pair like this would fetch from a wealthy collector. But could he get them both?

Carefully, he raised the shotgun to his shoulder, sighting along the long barrel to fix on the male bird. He hesitated momentarily before squeezing the trigger. Part of him felt shame for killing something so beautiful, but it was the only way to learn about the birds and to keep their memory alive as their numbers dwindled.

He fired the shot. One of the pellets hit the male

bird and it dropped to the ground. Instead of flying away, the female seemed to dither, startled by the noise and bobbing around the warm but lifeless body of her mate, poking him with her beak.

This time Buller had no hesitation. The huia mated for life, so the female bird would be likely to pine away anyway now that the male was dead. He quickly reloaded and the second bird fell alongside her mate.

Buller stood, finally able to stretch out his knees and back, and walked over to retrieve the warm corpses. The shot had not damaged their skins too much; they were still in saleable condition. It was a job well done.

The remarkable huia (*Heteralocha acutirostris*) was a magnet for nineteenth-century collectors. Its white-tipped tail and orange wattles were a striking contrast to its shiny coal-black plumage, and it was a scientific curiosity: the only bird species in the world where there was such a marked difference between the beaks of the male and female (a condition called sexual dimorphism). The male's bill was shorter and straight, the female's longer, and gracefully curved.

A forest-dwelling bird about the size of a magpie, the huia had been considered special long before Europeans laid eyes on them. Māori considered the huia a royal bird, its large black tail feathers, tipped with white, often being worn by chiefs. Sometimes only one or two feathers were worn in the chief's hair, but on special occasions important rangatira would wear a fan of all twelve of the huia's tail feathers, called a marereko. Sometimes the preserved head and curving bill of a female huia was worn as a pōhoi (ear ornament). Because the head is considered tapu (sacred) to Māori, and the chief's head was most tapu of all, items associated with it became tapu too. And so huia feathers, and the bird itself, were considered sacred.

Sometimes Māori wore huia feathers to a tangi, as a sign of respect for the dead. The feathers from the centre of the tail, where the white area passed straight across the feather, were considered more valuable than feathers from further out on the tail, where the white area fell on an angle. Because the feathers were so highly prized, and to keep them from being damaged, they were kept in special elaborately carved boxes called waka huia.

Māori used tari, long-handled snares, to catch the birds, calling them down from the trees by imitating the call from which their name comes—'hu-i-a'. The birds were then skinned and the feathered skins preserved. In pre-European times the birds were killed mainly for ceremonial purposes, although they were sometimes eaten. It was not until the late 1800s that large numbers of huia were killed by Māori, to satisfy European demand for specimens. However, it seems likely that the Pacific rat or kiore would have eaten the eggs of the low-nesting huia, and the bird's population may already have been in decline when European settlers started to arrive in Aotearoa in the early 1800s.

Huia were only ever found in the North Island, and they seemed to prefer to live on the open edges of forest rather than in the thick bush. Although fossil remains have been found all over the island, by the time European settlers arrived the huia mostly lived in the south of the North Island, from Te Urewera, the Central Plateau and the King Country down through the Manawatu to the Tararua, Remutaka and Ruahine ranges and on to Wellington, and east through the Wairarapa to Hawke's Bay.

Huia showed no fear and much curiosity towards humans, and they could easily be attracted by an imitation of their call. From the earliest European reports, huia were described as landing on people's hands or tolerating being picked up. They lived and fed in the understorey of the forest and on the ground. They were poor fliers and hopped about to gather their food of grubs and insects, frequently clinging to vertical tree trunks using their tails as extra grip, rather than perching on branches.

But the things which made the huia special also made them vulnerable. They were easy prey for the introduced predators which became widespread in the late 1800s, and they suffered from the destruction of their forest habitat. But another threat came from human collectors, whose fascination with the huia's uniqueness helped to bring about its doom.

Walter Buller was among those who took an early interest in the huia, although he has since been criticised for his collection of native species. Buller, the son of a Wesleyan missionary, was born in Northland in 1838, two years before the signing of the Treaty of Waitangi. When he was a child, his large family lived on the shores of Kaipara Harbour, where he spent his time exploring the bush and learning about its natural inhabitants, shooting birds to examine them more closely.

As an adult he became a lawyer and worked as a magistrate negotiating purchases of Māori land in the lower North Island. But his passion for birds persisted, and he continued to study and write about them. He is best known for his book *A History of the Birds of New Zealand*, first published in 1873 with illustrations by Johannes Gerardus Keulemans. Although Keulemans never visited New Zealand and painted the birds from mounted specimens only, these beautiful illustrations are today as famous as the book itself. They provide us with many familiar and iconic images of our vanished native wildlife.

In 1870, Buller presented a paper 'On the Structure and Habits of the Huia' to the members of the Wellington Philosophical Society. Although it was only a few decades since Europeans had begun colonising New Zealand, Buller said that the huia was already one of New Zealand's 'rarest and most valuable species', writing, 'Erelong it will exist only in our museums and other collections, and, [it is] for the sake of science that everything connected with its natural history should be faithfully recorded and preserved'.

Buller, like many others of his time, had a fatalistic attitude towards the extinction of native species. To his mind, it was the natural way of things for native animals and birds to die out and be replaced by the supposedly stronger, more advanced European species. (Buller's beliefs on this topic also extended to Māori people. He was not the only scientist or politician in the late 1800s who believed that New Zealand's original human inhabitants, as well as its birds, would pass away in the face of European settlement.) That's why, despite admiring the unique natural wonders of Aotearoa, he also thought it was acceptable to kill and collect them in the name of science, so that knowledge of them could be preserved even when the birds themselves were lost forever.

Buller had the odd pang of conscience, but by and large he believed he was doing the right thing by killing birds for preservation in museums and private collections. Describing an expedition with a Māori guide for huia in the northern Wairarapa in October 1883, he wrote:

> Rahui continued his call—a loud clear whistle . . . In a few seconds, without sound or warning of any kind, a Huia came bounding along, almost tumbling, through the close foliage of the pukapuka, and presented himself to view at such close range that it was impossible to fire. This gave me an opportunity of watching this beautiful bird and marking his noble bearing, if I may so express it, before I shot him . . .
>
> [Later that day] a pair of Huias, without uttering a sound, appeared in a tree overhead, and as they were caressing each other with their beautiful bills, a charge of No. 6 [shot] brought both to the ground together. The incident was rather touching, and I felt almost glad that the shot was not mine, although by no means loth [*sic*] to appropriate the two fine specimens.

Sixteen huia were shot dead on this expedition, one shot in the wing and wounded, and a further live pair were captured.

Literally thousands of birds were killed for sale to collectors in New Zealand and overseas in this way in the 1880s and 1890s. Buller recorded that in just one month in the early 1880s, a party of eleven Māori hunting on behalf of collectors 'scoured the wooded country lying between the Manawatu gorge and Akitio [on the Wairarapa coast], and brought in 646 skins; and a party of three men obtained a considerable number near Turakirai [Turakirae Head] on the south-western side of the Wairarapa Lake'. One collector alone, Austrian Andreas Reischek, took 212 pairs of huia for the Vienna Natural History Museum over a ten-year period from the late 1870s. However, Buller was quick to point the finger away from 'the zeal of collectors, who obtain large numbers for the European markets', writing, 'the natives annually kill a great many for the sake of their feathers'. From today's standpoint it seems unlikely that Māori traditional hunting had anything but the slightest impact on the huia in the face of the destruction of forests, introduced pests and European guns.

By the early 1890s it was clear that the huia was fast disappearing. In 1892, Buller spent several days in what had formerly been 'huia country' at the head of the Hutt Valley (Te Awakairangi) and saw only a single bird. (He shot it.) He also described the loss of the huia's habitat in the area known as Forty-Mile Bush in the northern Wairarapa:

> From Pahiatua we rode for twenty miles [32 kilometres] through clearings exhibiting nothing but charred and naked stumps, the

> whole of this country being at the time of my former visit covered with beautiful forest. From the practical standpoint of material advancement there is nothing regrettable in this; but the fact remains that the home of the Huia is being swept away, and, although these birds, in greatly-diminished numbers, have taken refuge in the wooded mountain-ranges, the date of their extinction cannot be very far distant.

As early as 1869, amateur naturalist Thomas Potts of Canterbury had written lamenting the loss of habitat and the decline of native species, and suggested the establishment of protective reserves. In 1885, Hugh Martin presented a paper to the Nelson Philosophical Society calling for areas to be set up to protect a list of more than 30 native birds, including the huia, the whēkau and the piopio (see chapter 4), the South Island kōkako (see chapter 7), the takahē (see chapter 11) and the kākāpō (see chapter 14). Martin also suggested paying a bounty for live specimens which could be moved to sanctuaries, noting that 'experience proves that no one will refrain from killing any rare or strange bird, unless it can be made in their interest to do so'. But although proposals such as this were discussed and debated, little actual action was taken.

One of the most influential advocates for the protection of the huia was Governor-General

William Hillier Onslow. The fourth Earl of Onslow was British-born (as all governors-general were until Sir Arthur Porritt was appointed to the role in 1967) and was sent out to the colony in 1889 to act as Queen Victoria's representative here. Onslow's second son was born in New Zealand the following year, and to commemorate the first child born to a governor-general in New Zealand and the fiftieth anniversary of the signing of the Treaty of Waitangi, the child was given the name Victor Alexander Herbert Huia. Known as Huia Onslow, his father presented him to the Ngāti Huia hapū (subtribe) of Ōtaki. In return, they asked him to request that the huia bird be protected by the government.

Onslow worked with none other than Walter Buller to draft a letter to Prime Minister John Ballance about the need for conservation of native species, the huia in particular. Onslow's last act before leaving New Zealand to return to England in 1892 was to sign into law an extension to the Wild Birds Protection Act, banning the killing or collecting of huia in any part of the country. (It is worth adding that Buller then went out and shot a few more, just to be sure he had enough specimens.) The bird was declared tapu by Māori leaders around the same time, and killing or collecting the huia in the Tararua Ranges was banned.

But it was too little, too late. The birds' population had already dwindled, and their slide towards extinction was gathering momentum. The Duke and Duchess of Cornwall visited New Zealand in June 1901, visiting Rotorua, where they were entertained by Māori performers and fireworks and presented with huia feathers to wear in their hair and hatbands.

This struck a further blow to huia. Non-Māori wearing huia feathers for fashion—which today would be considered an insensitive act of cultural appropriation—led to many of the remaining birds being poached for their feathers. (While killing the birds might have been illegal, selling their feathers was not.)

The last verified sighting of a huia came around 1905, in the Tararua Ranges north of Wellington, although reports of birds persisted into the 1920s. Onslow's pledge to respond to the plea of Ngāti Huia—'e hoa ma, puritia mai taku huia' (friends, hold on to my huia)—had come to nothing.

While there is no doubt that the felling of the great forests and the introduction of predators hastened the huia's demise, humans hunting and collecting the surviving birds did damage, too. Yet after all this destruction, little more than 100 mounted birds, skins and skeletons survive in museums in New Zealand, and a handful more in institutional backrooms and storage units around the world. No one knows how many more are held in private collections.

The huia was both exquisitely beautiful and utterly unique. Its image persists today as an icon of all we have lost and a reminder that we need to preserve what we have left. Perhaps in this way it can fulfil the whakataukī:

> *Huia e huia, tangata kotahi!*
> Huia, your destiny is to bring everyone together!

Lyall's wren (*Traversia lyalli*)

6. The light that went out: Lyall's wren

Martha Lyall wrestled the last item of clothing out of the tub and slopped it into the clothes basket. Now that there were three families living on the island and tending the lighthouse, it made for a lot of washing.

She kicked open the shed door and waddled down the shell path to the washing line, hip cocked to one side to rest the heavy basket on. Where were your children when you needed them to help? Out running wild in the bush, no doubt.

Martha dropped the basket to the ground and

reached into her pinny pocket for a couple of pegs. It was a good drying day, with a brisk wind blowing in off the sea. But then it was always windy on Stephens Island, at the very top of the South Island.

She could hear a strange noise in the shrubs behind her, as if something heavy was moving around. Martha turned to see one of the island's many wild cats—a stocky tabby, this one—jumping about in a pile of dead leaves at the base of a low-lying bush. She hissed at it through her teeth to shoo it away.

The cat didn't run away—it didn't even look up. It had something in its paws that it was playing with. Martha slumped a wet shirt back into the basket and went over to investigate.

She had thought it was a mouse, but once she got closer she could see that the cat's prize was a bird. It was only about the size of a mouse, though, and it was a mousy colour. It didn't seem to be making any attempt to fly away as the cat boxed at it with its paws. It wasn't much of a match for the cat, poor thing! Martha aimed a kick at the cat's mangy belly, and it hissed and dropped the bird before retreating into the shrubbery, where it continued to watch its victim.

Martha scooped up the little bird in her hands. Its eyes were half closed but its feathers were still warm, and it felt surprisingly soft.

'Can you fly, little bird?' Martha said softly. She tried to turn it upright on to its long, clawed legs, but it refused to stand up. It looked like she'd got to it too late.

She bent down and put it back on the ground. The cat might as well have it. She had work to do.

Stephens Island (Takapourewa) is a tiny outpost on the northwestern edge of the Marlborough Sounds, off the northern tip of Rangitoto ki te Tonga / D'Urville Island. Its rugged slopes were probably never permanently settled by Māori. The first Europeans to lay eyes on Takapourewa were Dutchman Abel Tasman and his crew, who sheltered nearby at Christmas 1642. Captain James Cook, whose ship *Resolution* was nearly caught in a nearby waterspout in 1773, named the island after Sir Philip Stephens, Secretary of the Admiralty.

It is a rocky place, edged with sheer cliffs that rise up out of the frequently turbulent sea. The island is only 1.5 square kilometres but stands very high, rising to a summit of 283 metres, which is just taller than Rangitoto in Auckland's Hauraki Gulf / Tīkapa Moana. The first Europeans to land on Takapourewa found it covered in thick, hardy, wind-blasted forest—and absolutely alive with birds.

In 1892, teams of workers travelled to the island to construct a lighthouse on its inhospitable slopes. Officials had wanted to build one there since the 1850s, as shipping increased between the growing colonies of Wellington and Nelson, and New Zealand and Australia. However, Takapourewa was so rugged it was hard to even climb it to work out where to put a lighthouse. It was not until 1891 that the island was taken by the government from its owners, Ngāti Koata, for the purposes of building a light. A lantern, a lens and other machinery for the light were ordered, and the kitset iron tower was manufactured in Auckland. A wharf and tramway were built to get supplies ashore and up to the building site, nearly 200 metres above sea level. The horse that pulled the materials up in a trolley even had to haul its own feed, as there was no grass on the island for it to eat.

At this time, Takapourewa was covered in pure, untouched native vegetation—mostly hardy low-growing shrubs and trees which could withstand the salty winds of Cook Strait (Raukawa Moana)—and populated by literally thousands of birds. They were so numerous and unafraid of people that the lighthouse builders had to shoo them away while they were cooking outside over fires, or else the birds would land on the pots. The South Island variety of the piopio (see chapter 4) and tīeke (saddleback) were abundant, despite both being rare on the mainland by this time, and the air was thick with the chattering of kākā and kākāriki and the songs of tūī and korimako (bellbird).

Takapourewa was also crawling with a strange lizard that was found on other offshore islands around New Zealand: the mysterious and ancient tuatara. Elsewhere they were hard to find and had been little studied: here, on an undisturbed island rich in the seabird chicks, lizards and invertebrates tuatara eat, they were so thick on the ground that the workers had to take care to avoid stepping on them.

There was one particularly special bird on the island. This little wren, which would become known as the Stephens Island or Lyall's wren (*Traversia lyalli*), was the smallest flightless bird in the world. No bigger than a mouse and a speckled brown colour, it seemed to come out at dusk, when it would quickly run around on the ground and hide among the rocks. Fossils show that this type of wren once lived all over the North and South islands, but introduced predators had wiped it out on the mainland. Stephens Island was the only place in the world it was still found when the lighthouse keepers came.

In January 1894, the new light beamed out for the first time. It was operated by three keepers—Patrick Henaghan, David Lyall and William Arnold—who moved to the isolated island with their families and a teacher for their children. They cleared some of the island's vegetation to create farmland, and they brought over chickens, cows and sheep for eggs, milk and meat. It was a rugged and isolated existence as they had infrequent contact with the outside world and only the birds and each other for company.

As well as the supply ship, which came every three or four months bringing much-needed fresh provisions, the island occasionally received other visitors. Among those who came were natural-history enthusiasts interested in seeing the island's unique wildlife for themselves.

Amateur naturalist and collector Edward Lukins of Nelson spent a day and a night there in October 1894, arriving on the supply boat along with nine live sheep. Nine months after the arrival of the lighthouse keepers, Lukins made a list of 31 different types of bird on the island, which included the 'rock wren'. He noted, 'The native thrush, now almost extinct on the main land, is so numerous that there was scarcely a bush in which at least one could not be seen, and they could be caught without difficulty. The saddle back is almost as numerous there, rare though it generally is now.' On his return, he wrote in Nelson newspaper *The Colonist*:

> It was with considerable interest I found so many of the more rare of the New Zealand birds on Stephens Island and I hope they may long continue to thrive on this inaccessible island where they will certainly be unmolested by stoats and weasels.

He didn't think to mention cats.

No one knows for sure when the first cats arrived on Takapourewa. The lighthouse keepers may have brought domestic cats with them, although the island had no rats or mice for them to hunt. Another story goes that a pet cat was brought ashore from the supply boat *Hinemoa* in a sack, and one of the children living on the island accidentally let it out and it ran off and became wild. Either way, soon after the arrival of the first permanent human inhabitants, cats were making short work of the island's unique species.

Cats were the reason the island's unique flightless wren came to the attention of the world in the first place. Not long after he arrived on the island, keeper David Lyall started skinning and preserving birds killed by the island's cats. Lyall gave one of the skins to Alexander Bethune, the second mate of the *Hinemoa*, to be passed on to naturalist Walter Buller, who was then considered New Zealand's foremost expert on native birds. Buller was very excited to 'discover' a new species, and he sent it off to London to be classified and described. He wrote in 1895:

> There is probably nothing so refreshing to the soul of a naturalist as the discovery of a new species. Quite apart from the satisfaction of being able to impose a specific name which, according to the accepted rules of zoological nomenclature [the naming of animal species],

must be respected for all time, there is an indescribable charm in the mere fact of discovery.

In fact, in this instance Buller had been gazumped. Lyall had also been selling skins to Henry Travers, a collector and trader in bird specimens from Wellington. Travers was an astute businessman, who had visited the island in the early 1890s and heard about the unusual flightless wren. Although Travers sometimes supplied specimens to Buller, when he received the skins of nine dead wrens in 1894 he sold them to a higher bidder, the wealthy British banker and dabbler in natural history Walter Rothschild.

Lord Rothschild was an eccentric man who used his inherited wealth to stock his own private zoo and museum at Tring Park, northwest of London. He was also a client of Buller's, but he showed no mercy when it came to one-upping his supplier. He wrote a paper describing the new species for the British Ornithologists' Club, and had it published before Buller's report. The name given to the species by Rothschild—*Traversia lyalli,* after both Travers and the lighthouse keeper Lyall—was then adopted over Buller's suggestion of *Xenicus insularis,* much to Buller's displeasure.

But almost as soon as the wren's existence was revealed to the wider world, its disappearance was, too. In March 1895, *The Press* in Christchurch ran a report of Rothschild's naming of the wren, under the heading 'Found and exterminated'.

> ... there is very good reason to believe that the bird is no longer to be found on the island, and, as it is not known to exist anywhere else, it has apparently become quite extinct. This is probably a record performance in the way of extermination. The English scientific world will hear almost simultaneously of the bird's discovery and its disappearance before anything is known of its life-history or its habits.

It was 'saddening to reflect', the author wrote, 'how, one by one, the rare and wonderful birds which have made New Zealand an object of supreme interest among scientists all over the world are gradually becoming extinct'. The article finished by advising the Marine Department to ban lighthouse keepers from having cats, 'even if mouse-traps have to be furnished at the cost of the State'.

Travers returned to the island in early 1895, but could not find a single wren. By the end of the year, he reported to Rothschild's museum curator that even Lyall now believed the bird 'to be quite extinct as even the thrushes and saddle backs which were at one time very numerous are now almost of the past'. Travers still had a few more specimens up his sleeve, which he sold over time, but there was no further sign of live wrens.

Overall, just fifteen specimens of this unique and unusual bird were ever collected. Today, they are scattered among museums around the world, with only three in New Zealand. Yet another of the wonders of Aotearoa had been extinguished, this time by cats.

Lyall's name remains attached to the wren forever more—not just because of its official and scientific titles, but also because the legend of 'the lighthouse keeper's cat' has featured widely in the telling of this extinction story. But it wasn't just one cat—and probably not a cat belonging to the Lyalls—that did the damage. In fact, former wildlife officer Derek Brown, in his history of the island, points out that if it hadn't been for Lyall collecting his specimens, the bird probably would have become extinct before anyone even knew it existed. We might know next to nothing about the bird and its habits, but at least we know it was there.

It wasn't until a few years after the disappearance of the wren—along with the piopio and many other birds—from the island that something was done about the cats. In 1897, principal lighthouse keeper Patrick Henaghan requested shotguns and a supply of ammunition to try to get them under control. His replacement, Robert Cathcart, shot more than 100 wild cats in eight months in 1899. By early 1900, he and his assistant keepers had run out of shot.

In the four years from the start of 1902, keepers recorded shooting more than 400 cats, 300 of them in 1905 alone. By the end of that year, however, numbers had dwindled, and the population never recovered. By the 1920s it was believed that, like the birds they had once preyed on, all the cats were gone.

The removal of the cats might have made it possible for what was left of the island's bird, insect, reptile and invertebrate life to recover, but felling of the bush continued. By 1918 only a few patches of trees remained amid a landscape of windblown pasture. The island continued to be farmed until 1966, when it was finally declared a wildlife sanctuary—nearly 70 years after most of its precious species had been eradicated.

In 1895 a case had been made for Stephens Island to become a reserve, mainly to protect from collectors the thousand or so tuatara still found on the island. But instead of setting aside the island, the government chose to legally protect the tuatara instead.

The last lighthouse keepers left in 1989, when the light was fully automated. The island was returned to Ngāti Koata in 1994 as part of a Treaty of Waitangi settlement, but the iwi gifted it back to the government to be managed as a sanctuary. Today the original keepers' houses are occupied by Department of Conservation (DOC) staff and volunteers working on the ecological restoration of the island.

As a closed sanctuary, the still predator-free island is now host to the world's largest population of tuatara—more than 30,000 of them. They share the island with seven different types of rare lizard, large colonies of seabirds and a population of the fearsome-looking but harmless Cook Strait giant wētā. Takapourewa is also one of the last homes of the highly endangered native Hamilton's frog, whose population is carefully managed by DOC staff (and kept away from the tuatara, who occasionally eat them). Only about 300 of these primitive frogs survive, and their small populations make them highly vulnerable to introduced pests and diseases.

Every night, as it has for more than a century, the beam of the Stephens Island lighthouse continues to sweep across the sea. But the light of Lyall's wren has been extinguished for good.

South Island kōkako (*Callaeas cinerea*)

7. In search of the grey ghost:
South Island kōkako

Two trampers were walking along a section of the Heaphy Track in the Kahurangi National Park at the top of the South Island. It had rained all night but the day had dawned fine, so they packed up early and set off along the track towards the next hut. It was about 20 kilometres away—six hours' walking would do it.

They were in remote country, where northwest Nelson meets the wild West Coast. Far from towns and farmland, the bush is dense and lush. Some of it has never been logged, and is as close to the primeval forest that once cloaked Aotearoa as you can get. It's strangely subtropical for so far south, with giant nīkau palms and patches of glossy broadleaf trees among the beech.

As they walked, boots squishing on the damp ground, the trampers talked about the bird song they had heard earlier that morning. Just before first light, the dawn chorus had started up. First there were a few mellifluous notes of tūī, a pair communicating with each other in call and response. Then came the high-pitched twittering of fantails and bush robins. Swooping notes from kererū, and the ringing of korimako. Finally, away in the distance, a long, strong low note like chords being played on a distant organ—deeper than the vocal gymnastics of a tūī, richer and more sonorous than a bellbird. Unique. Remarkable. They wondered what they could have heard.

The trampers were starting to drop down towards the Heaphy River when a movement in the forest caught one tramper's eye. Among the lower branches of the beech trees nearby, she had a fleeting glimpse of a large grey bird.

'Stop!' She put one hand on her companion's arm and pointed into the trees. 'Look.'

On the downslope below them, the tree canopy was at eye level. About 5 metres away, a grey bird a bit smaller than a magpie was hopping heavily along a branch. It moved from bough to bough by jumping, not flying. They could see a flash of orange colour at its bill, a dark mask like a bandit around its face. Once, just once, it opened its beak and let drop a low, sustained, liquid note so beautiful it was like a jewel hanging in the forest air.

'Do you think . . .' she said, at the same time her partner said, 'It can't be . . .'

They would never find out. At the sound of their hushed voices, the bird took one large hop, spread its wings and glided further off into the trees. They did not see or hear it again.

This is the story of a bird which may or may not be extinct. If you are a member of the scientific community, you would probably say it is gone forever. However, if you are one of the many back-country travellers who have heard or seen the bird in the past 50 years, you might disagree.

Until recently, the last accepted sighting of the South Island kōkako (*Callaeas cinerea*) was in the Mount Aspiring National Park in 1967. The species was officially declared extinct by the Department of Conservation 41 years later, in 2008. However, there are a small but significant number of conservationists who believe that tiny remnant populations still exist, and they are on a mission to prove it. In 2012, a mounting number of sightings of what were believed to be South Island kōkako led to its reclassification in the enigmatic 'data deficient' category—which means that we simply do not have enough evidence to decide one way or the other. The most recent accepted sighting is now from 2007, at Rainy Creek near Reefton on the West Coast.

Once (and maybe still now), two species of kōkako lived in Aotearoa. The two types could be told apart by where they lived—in the North or the South Island—and also by the colour of their wattles, the fleshy flaps at the base of their beak. In the North Island the species has blue wattles, while the bird found in the South Island had wattles that ranged in colour from pale yellow to orange or red. To confuse matters, there are a few reports of North Island kōkako with orange wattles and some of the South Island birds with blue wattles.

According to Māori legend, Māui called out to the birds to bring him water as he fought to slow the sun's path across the sky. The kōkako heeded his call, carrying water in its wattles. Māui rewarded the bird by giving it long, slender legs so it could bound through the forest. Even in pre-European times when the birds were still abundant, they were known to be hard to catch. One pepeha says 'E hoki i kona, e kore e mau i a koe te kōkako a Whareatua' (You can turn back, for you will never catch the kōkako of Whareatua). Māori would blow a pehe—a 'call leaf' used to imitate the sounds of birds—to attract them. However, kōkako were seldom eaten by Māori. Early ethnologist Elsdon Best wrote, 'as one old man put it to me it bears too close a resemblance to the shag; this in reference to its lack of palatableness [tastiness]'.

Like its North Island cousin, the southern kōkako was quite a large bird, a little bigger than a rock pigeon, with dusky grey-blue plumage and a striking black mask around its eyes that gave it the air of a bandit. It had small wings for its body size and was not a great flier. Perhaps the most striking feature of the kōkako was its song. Fortunately, through the hard work of conservationists who saved the North Island species, we can still hear what its melodious voice must have sounded like: strong, rich, bell-like tones that resonate like a metal tube being struck, chiming through the forest at dawn and dusk.

The South Island kōkako was observed and described by the first European voyagers to come ashore in Aotearoa. Sydney Parkinson, one of the artists aboard Captain James Cook's *Endeavour* in 1769–70, recorded seeing three types of 'wattle birds' at Meretoto / Ship Cove in Queen Charlotte Sound / Tōtaranui—the others were probably both tīeke (the South Island saddleback, *Philesturnus carunculatus*), which has a black and rusty-orange body, but is chocolate brown all over as a young bird. At least one dead kōkako from Meretoto was taken back to England—on his return home, Joseph Banks, the leading gentleman naturalist on the expedition, gave a skin to his friend, ornithologist and collector Marmaduke Tunstall.

More birds were seen—and shot—on Cook's second voyage in 1772–75. In autumn 1773, the crew of the *Resolution* visited Tamatea in the southwest of the South Island, which Cook named Dusky Sound, and spent more than a month resting and stocking up on food and water after their voyage through the Southern Ocean. Naturalist George Forster, who travelled aboard the *Resolution* with his father Johann Reinhold Forster, mentions the kōkako in his journals. He shot one at Cascade Cove, and more specimens were killed and collected when the ship coasted north to Cook's favoured anchorage at Meretoto.

Back in Europe, where beautiful things from the other side of the world were named and classified, the South Island kōkako was given the Latin name *Glaucopis cinerea* by German naturalist Johann Friedrich Gmelin in 1788. The first part of the name means 'grey-eyed' and the second part refers to the ashy colour of the bird's feathers. (Its Latin name has now been changed to *Callaeas cinereus*, and the North Island bird is recognised as a separate species, *Callaeas wilsoni*.) The South Island bird was also known to Europeans by the ugly and unimaginative name of orange-wattled crow.

The bird appeared abundant to Cook and other early visitors, but once European settlement began in earnest the kōkako began to retreat along with the forest. By 1873, naturalist Thomas Potts wrote that the bird was:

> being driven away by the approach of the colonist, for as the coast-line of a large portion of New Zealand exhibited signs, or echoed the sounds of the work of the settler in his encroachments on the tangled wilderness of nature, the Kōkako retired to the higher and more remote bushes of the interior . . . It is not a matter for surprise that the Wattle-bird is no longer to be found in its old haunts; it seeks shelter amongst the higher parts of the bushy gullies—a refuge at once precarious and temporary.

Potts was one of the first of the European settlers to express conservationist sentiments, pointing out the damage done by loss of habitat and the introduction of new species to compete with native ones. He lobbied for the establishment of reserves to protect native species, including on Resolution Island in Dusky Sound, although this did not become a sanctuary until 1891, several years after Potts's death.

The knowledge that the birds were declining did not stop collectors from shooting them to create stuffed specimens. Austrian taxidermist Andreas Reischek wrote in 1885 of shooting one of a pair of South Island kōkako, then lying in wait for its mate to appear so he could bag them both. However,

> . . . to my astonishment, instead of flying away

> when it saw me, the poor thing went to its dead companion, hopping around and calling, evidently in a great state of agitation. I felt so much for this bird, that I was very sorry I had shot its mate, and let it go.

Kōkako are poor fliers, preferring to bound from branch to branch and glide short distances. Potts described the southern bird's habit of feeding near the ground, eating leaves and berries and foraging for tasty plant matter and small insects among moss on tree trunks and fallen branches. Bushman Charles Douglas, known as Mr Explorer Douglas, described the birds' habits:

> They prefer to run about the limbs and trunks of the low scrub, they never actually walk, but run with a strolling sort of a gait that is very funny. They are a harmless playfull [*sic*] sort of bird, easy to tame and can be caught either with a snare, or by applying salt to their tails.

He also described the birds' cry as 'indescribably mournful . . . and sadly suggestive of departed spirits'. In the deep forest, however,

> they can be heard to perfection. Their notes are very few, but are the sweetest and most mellow-toned I ever heard a bird produce. When singing,

> they cast their eyes upwards . . . and pour forth three or four notes, softer and sweeter than an aeolian harp or a well toned clarionet [*sic*].

Walter Buller was also taken by their call, noting that 'it is well worth a night's discomfort in the bush to be awakened at dawn by this rare forest music'.

Fortunately, due to the actions of conservationists in the 1970s and 80s, the world is still blessed with the song of the North Island kōkako. At this time numbers of the blue-wattled bird were falling, and one of their last strongholds, Pureora Forest to the west of Lake Taupō (Taupomoana), was still being felled and even burned to clear land for pine plantations. Activists scaled the giant tōtara trees and camped up them to stop them from being chopped down. The protests caught the eye of the media, the government was forced to halt logging, and this vital piece of habitat was saved.

The protest also raised the profile of the largely forgotten kōkako, and for the first time money was put towards developing plans for its conservation. With the protection of forestland, predator control, captive breeding and translocation (moving birds) between sanctuary sites, the North Island kōkako was pulled back from the brink of extinction. New populations were established on Te Hauturu-o-Toi / Little Barrier Island and Kapiti Island in the 1980s, and on Tiritiri Matangi in the 1990s. There are also now groups of birds on the mainland in sanctuaries such as Maungatautari in the Waikato and Boundary Stream Mainland Island in Hawke's Bay, as well as in forests in Northland, Taranaki and the Bay of Plenty, in Te Urewera, and in Pureora and Whirinaki (another forest saved from the bushman's axe through direct-action conservation in the 1980s).

Fewer than 400 pairs of North Island kōkako were estimated to exist in the late 1990s, when a management plan was written with a twenty-year vision of restoring the national population to around 1000 pairs by the year 2020. As of late 2019, there are estimated to be 1600 pairs of North Island kōkako singing their songs in sanctuaries and reserves, including in the Hunua and Waitākere ranges on the edges of the city of Auckland.

The South Island kōkako's story is not so happy. The birds became more and more rare towards the end of the nineteenth century. In 1888, Walter Buller described its distribution as 'irregular'—in fact, it was vanishing. Kōkako are territorial, which means that each pair 'marks off' their own large patch of bush through song, so the population suffered as the large areas of forest they needed to survive were felled and fragmented. The southern kōkako's habit of feeding and nesting close to the ground made it highly vulnerable to introduced predators. Female kōkako incubate their eggs (sit on them to keep them warm) on their own for nearly three weeks, so they were especially at risk of being killed by rats.

Fewer and fewer sightings were recorded as the birds retreated into the deep bush, seemingly hiding from human view. The species stumbled on, largely unseen, for the first half of the twentieth century. In 1967, near Makarora on the edge of the Mount Aspiring National Park, a musterer stopped to listen to the roar of a nearby river when a kōkako appeared,

> walking along a log which protruded from a thick patch of fern . . . I think it saw me immediately because it quickened its pace, flew from the end of the log to a sloping tree trunk a short distance below, and began to climb the trunk in a most peculiar way. With each rather ungainly step upwards, it appeared to hold on to the bark with its beak, look in my direction, take another step, hold, look, and so on until it reached the branches, when it hopped rapidly out of sight.

And with that, the species disappeared from view.

The South Island kōkako was officially declared extinct in 2008. However, stories of the bird's continued existence persisted. As late as the 1990s and early 2000s, people regularly reported seeing the birds or hearing their calls, particularly in the west and northwest of the South Island, Marlborough and the Catlins, on the southeast coast of the South Island. A number of reports came from the aptly named Rainy Creek, inland from Reefton on the West Coast and in March 2007 two observers provided sufficient detail of a sighting there to convince authorities that the bird was not extinct.

Due to this report, the South Island kōkako was reclassified as 'data deficient' in 2013. But, despite the South Island Kōkako Charitable Trust offering a $10,000 reward for photographic or physical evidence that the bird survives, the grey ghost remains elusive.

Rhys Buckingham has spent many years searching for the South Island kōkako. He had his first encounters with the 'extinct' bird more than 40 years ago, when he heard its calls—a 'bonging' sound 'like cathedral bells'—near Lake Monowai in Fiordland in the late 1970s. He heard the calls again near Lake Wakatipu and on Rakiura / Stewart Island in the 1980s, when he was part of expeditions to try to find kōkako.

Buckingham says he is sure that kōkako were still present on Rakiura then, but the birds did not want to reveal themselves. His team came close: they heard a large number of calls, some at very close range; they found moss that had been freshly 'grubbed' by the birds; and they had a handful of brief sightings. Buckingham says that hearing the calls 'was like stepping back a century—we were hearing calls that were just not heard in any other forest in New Zealand at that time. It was so exciting.' Buckingham and his co-workers also heard birds, but never saw them, in the Greenstone and Caples valleys in Fiordland in the 1980s. Unfortunately, the search was eventually abandoned without producing any hard results.

But Buckingham has gone on looking and hoping. He still works in the field, following up reports and trying to track down any last survivors of this unique species. The journey has been a frustrating one at times, a tale of missed opportunities and what-ifs. A clear recording made of a kōkako call in Westland was lost in a house fire. A feather found on Rakiura was taken overseas and not returned. Cameras and recording devices were not at hand at critical times. The birds seem to 'throw' their voices and hide in foliage. It is like they do not want to be found.

Although reports keep coming in from trampers and people working in the back country, Buckingham thinks the South Island kōkako may be functionally extinct—that is, only a few lone male birds survive, so the species cannot breed. The question you have to ask, Buckingham says, is how a few scattered birds could possibly have survived for so long, when they were already rare 130 years ago.

Buckingham says many of the recent sightings and call reports he has investigated involve trying to work out why the birds reported appear to have such a cryptic habit—avoiding being seen—to be certain he is not chasing tūī and other clever mimics that might have learned and remembered kōkako calls.

'Recent observations suggest that the combination of odd calls and super-cryptic behaviour are indeed more likely to be from a kōkako, because tūī and their kin do not display this kind of behaviour. Indeed, the fact that tūī and bellbirds readily and predictably copy the odd calls add further evidence that South Island kōkako are indeed still out there, in rare, isolated "populations",' Buckingham says.

'Maybe they are not even "functionally extinct". They are just so elusive that it is impossible to estimate numbers or whether they are paired or not. [The challenge is] how can we catch such a bird to learn about it and save it through translocation?'

Buckingham is optimistic that the grey ghost will yet be found. It will be a happy day for its many believers when that happens.

South Island snipe (*Coencorypha iredalei*)

8. The tragedy of Big South Cape:

South Island snipe, Stead's bush wren, greater short-tailed bat

It was a relief to be back on shore. The boat trip round the windswept coast of Rakiura had taken longer than usual and Jimmy was happy to have his feet back on dry land. He shouldered his bag of provisions and started heading up the track towards the small house he shared with his whānau over the muttonbirding season.

Walking up the slope, he took a deep breath and filled his lungs with the special smell of Taukihepa—the air smelled sweeter, somehow, than on the mainland, more pure and natural. There wasn't as much birdsong as usual, but maybe that was because the sky was heavily overcast. Usually the island was alive with birds, which kept the workers company as they processed the muttonbirds by day. It wasn't uncommon to be chattered to by a flock of tīeke, or to find one curled up asleep in your jacket pocket if you left it lying in the sun. Tūī, korimako, kākā, kākāriki, mātuhi, kakaruwai—the island was alive with birds.

Jimmy caught a flash of movement out of the corner of his eye. He turned and saw that he was being watched. Sitting on the track behind him, bold as you like, was the biggest rat he had ever seen. It must have been nearly two foot long, from the tip of its twitching nose to the end of its ropey tail. It looked fat, too, as if it had been having a good feed. It stared back at him, wiping its translucent ears with a tiny pink hand before scampering back into the bush.

Rats. That was no good. Jimmy had heard from fishermen that there were a few around, mostly up the other end of the island, and he didn't fancy having to get rid of them.

He was almost at the house when he heard a shout from inside.

'Aiii! Look at the mess in here!'

Jimmy dropped his load on the ground and ran

up the steps. Inside, his uncle was standing with his hands on his hips, looking around wide-mouthed at a scene of utter destruction.

There were rats here, all right. They'd eaten their way into the supplies left carefully stowed the previous year. Flour was scattered all over the floor, along with chunks of nibbled potato. But the rats hadn't stopped at food. Newspapers had been torn to shreds. The fluffy kapok insides of pillows and mattresses were strewn everywhere, where the rats had made their nests. The little blighters had even stripped off the wallpaper, trying to get at the tasty paste. Anything they hadn't eaten they'd soiled with their droppings and urine. It was a mess.

Even as Jimmy and his uncle watched, another fat rat stuck its head inside and stared them in the eye. It seemed to be laughing at them.

Once a year, in late March, muttonbirders head to Taukihepa / Big South Cape Island. This rocky-shored, bush-covered outpost of land lies about 1.5 kilometres off the southwestern tip of Rakiura / Stewart Island. At 10.4 square kilometres, Taukihepa is the largest of the 36 traditional muttonbirding islands, retained in collective Māori ownership through generations. Each year, Rakiura whānau make the arduous journey to the edge of the Southern Ocean to harvest the fledgling (young) chicks of tītī, the oily seabirds also known as the sooty shearwater (*Puffinus griseus*) or muttonbirds. The chubby chicks are extracted from their burrows or caught above ground by torchlight, and they are plucked and preserved. Only families with traditional rights can visit the islands to harvest birds on their piece of land, or manu, and birds can only be taken during the two-month season which starts on 1 April each year.

When the muttonbirders arrived on Taukihepa in early 1964, they immediately noticed that something was different. For a start, their houses and buildings had been ransacked by a rat population which seemed to have exploded since the previous year. They also noticed that the birds were gone—not their precious tītī, which seemed largely unaffected by the plague of rats, but most of the other native species which normally filled the air with their song. Something was seriously wrong, and ship rats were to blame.

Because of its extreme isolation, even by the mid-twentieth century Taukihepa was still virtually untouched, a last, shining example of what most of New Zealand must once have been like. Apart from the annual visits of the muttonbird families, it had never been occupied by humans, and it was still covered in thick, weather-resistant forest. Among the range of species living there in the early 1960s were three special types of bird found nowhere else in New Zealand, and a bat no one knew was endangered.

The tīeke, or saddleback, had once been a common

sight on the mainland. Fearless and inquisitive, the birds have handsome black feathers with a marked chestnut-coloured 'saddle', and small orange wattles like their relatives the kōkako. Unlike the North Island variant *Philesturnus rufusater*, the young of the South Island species (*Philesturnus carunculatus*) are chocolate brown all over for the first year of their lives. (Initial observers, including ornithologist Walter Buller, thought they were a separate species and called them jackbirds.)

Māori legend has it that the bird got its distinctive saddle at the hand of Māui. It refused to fetch him water during his epic battle with the sun, so he scorched its back with his fiery hand. (The kōkako was more helpful and fared better—see chapter 7.) Because tīeke roosted and nested close to the ground, and had no fear of human or animal predators, they were easy pickings for rats, stoats and weasels, and quickly disappeared from the mainland when these pests spread during the second half of the nineteenth century. By the 1930s, tīeke survived on just a few offshore islands around the coast: the northern species on Taranga (Hen) Island in the northern Hauraki Gulf / Tīkapa Moana, and the southern type on Taukihepa and two small islands nearby.

Taukihepa was also the sole remaining home of another, less showy type of bird: Stead's bush wren (*Xenicus longipes variabilis*). Bush wrens were tiny, less than 10 centimetres long and weighing just 16 grams, not much more than a cherry tomato. Their feathers were greeny-brown, with a white stripe over their eyes and a grey chest. They had long legs and toes and moved quickly, bobbing up and down when perched and flying over short distances.

Elsewhere in New Zealand there were once two other subspecies of bush wren, mātuhi or mātuhituhi to the Māori, also now extinct: the North Island bush wren (*Xenicus longipes stokesii*), last seen at Lake Waikaremoana in 1955, and the South Island bush wren (*Xenicus longipes longipes*), which lingered on a little longer than its relatives and was last recorded in 1968 near the Nelson Lakes.

In 1925, pioneer naturalist Herbert Guthrie-Smith had been the first European to describe the southern birds, then in 1935 ornithologist Edgar Stead recognised this wren as a distinct subspecies. Stead's bush wren had originally also been found on Rakiura, but by 1963 it was clinging on to its last refuge, Taukihepa.

The third unique species was less seldom seen but more frequently heard. Southern Māori told of the legendary hakawai, a giant bird which swooped from above on to its prey. While part of that story may come from memories of the extinct giant eagle (see chapter 2), the fearsome sound of the hakawai could still be heard at night on Taukihepa. The early whalers called these noisy birds 'Breaksea Devils', after one of the islands where the sound was heard. Surveyor Frederick Tuckett wrote in his diary in 1844:

> All the people frequenting this coast [Foveaux Strait] believe in the existance [*sic*] of an extraordinary bird, or phantom, which they can never see, but only hear rushing past them with the rapidity of a falling rocket, and making a terrible rushing sound. The Maories [*sic*] declare that it is a bird possessing many joints in its wings.

But the bird that was making the noise on Taukihepa was not large or frightening at all. Instead, the screeching, tearing noise—described by some muttonbirders as sounding like an anchor chain being lowered, or the jingling of chains being drawn through space—was made by a shy, mottled-brown bird called the tutukiwi or South Island snipe (*Coenocorypha iredalei*). About the size of a blackbird, the snipe was largely quiet during the day, but on moonlit nights it would fly high then plunge down out of the sky with an eldritch roar, its tail feathers vibrating to create a unique and disturbing noise.

There was a fourth species unique to Taukihepa. It could also fly, but it wasn't a bird. It was something even more rare and unusual—one of the only land mammals to be found on the islands of Aotearoa. At the time of the rat invasion, no one knew that the island was the last home of the greater short-tailed bat

(*Mystacina robusta*). These furry, pig-faced creatures, which hung upside down in rocky caves during the day and fluttered around the muttonbirders at night, weren't declared a separate species to the short-tailed bat found on the mainland and on nearby Rakiura until long after they had become extinct.

Mystacina bats are ancient creatures unlike other bats around the world in that they hunt on the ground, rather than in the air. These tiny pekapeka come out at night to rootle around on the forest floor in search of insects and fruit, using their folded wings as 'front limbs' to crawl about. The bats of Taukihepa weighed just 12–15 grams—about the same as a strawberry—and their furry bodies were grey, like mice, with large ears and little beady eyes.

During the nineteenth and early twentieth century, Taukihepa was visited only by muttonbirders, fishermen and the occasional naturalist. Guthrie-Smith, who was familiar with the damage done to the mainland environment by man and introduced animals, declared the island 'an ark for native bird species'. Sadly, he also foresaw the island's future, writing 'always hangs overhead the sword of Damocles; should rats obtain a footing, farewell to Snipe, Robin, Bush Wren and Saddleback'.

And so it came to be. In 1964, once the mutton-birders had returned to the mainland they reported the rat infestation to the New Zealand Wildlife Service. Two officers were sent to Taukihepa and found the situation to be critical. Not only had the rats devastated the bird population, eating eggs, chicks and even small adult birds, they had also decimated the numbers of insects, invertebrates and reptiles such as geckos which once lived on the forest floor and in the trees. They had gnawed many of the smaller shrubs back to bare twigs. The island was dying.

Something had to be done. But those who wanted to act faced opposition even from within the ranks of the Wildlife Service. Some people refused to believe that rats alone could possibly be responsible for the loss of these species. Other scientists believed that the rats and the birds would reach a 'natural balance', where numbers of both would stablise.

The Wildlife Service developed a plan: as well as laying poison baits to try to control the rats, they would send a team to catch any of the remaining snipe, wrens and saddlebacks and try to move them to a safer place. Fortunately, the service had recently been working out ways to catch and transport North Island saddlebacks from Taranga to other predator-free islands to make the northern species less vulnerable to being wiped out. Don Merton, a young Wildlife Service officer, and his co-workers had developed a way of luring the birds using tape recordings of their songs and catching them in a 'mist' net with very fine mesh all but invisible to the birds.

A small team, including Merton and another officer, Brian Bell, were dispatched to the island on a New Zealand Navy ship. As winter set in, they battled gales and freezing temperatures, hail and snow, to investigate nearby islands to which the birds could be transferred. On Taukihepa they set up their base and built aviaries in which to keep the live birds once they had been caught.

Using the techniques learnt on Taranga, 36 saddlebacks were captured and transferred to the nearby rat-free Kaimohu Island and Big Island. But catching the secretive snipe and the tiny wrens proved much harder. In the 1983 *Wild South* documentary 'Island Eaten By Rats', which told the story of the Big South Cape tragedy, Merton said that catching the snipe

> was a little like catching butterflies. We would advance across an open area and flush a snipe. Then we would sneak up to where it landed with our hand-nets, and attempt to catch it as it tried to fly off a short distance once more.

The team found and caught only three snipe, but two of these died before transfer and the third managed to escape. That was the end of the South Island snipe.

The wrens were almost as tricky. They had to be scooped off the mist nets on the ground before they could climb through the holes and escape back into the undergrowth. Nine were captured, but they proved hard to keep alive in captivity. Although the Wildlife Service team hand-fed them with insects and honey-water, three died soon after being caught. In the end, six survived to be transferred to Kaimohu Island.

The sea was running high when the naval patrol boat took its precious cargo to the new sanctuary. It was too rough to land on the island's rocky shore in a dinghy, so the boat was carefully nosed in to a shelving rock, and Bell jumped ashore. The birds in their boxes were winched across, and Bell set them free before jumping into the sea and being dragged on a rope back through the icy waters to the boat. They had done all they could.

The following summer, the wildlife officers returned to the southern islands to find the saddlebacks doing well. The bush wrens, however, had failed to breed. Over the next few years the last birds died off. Despite the best efforts of Bell, Merton and their team, Stead's bush wren was now extinct.

That same year, the Wildlife Service team recorded the presence of a bat colony in a cave near Puawai Beach on Taukihepa, although no bats were seen, just lots of dung with 'a distinctive smell'. After a few more sightings in the late 1960s, no further animals were found. Saving the bats was not considered a priority as they were not at the time thought to be a unique species, and so another of the wonders of Aotearoa was lost.

However, in the late 1990s, bat calls were recorded on nearby Putauhina Island, and since then there have been several potential sightings of bats there and on Taukihepa. No one is sure whether any greater short-tailed bats are alive today, but the species has been named one of the world's 25 'most wanted' species by the organisation Global Wildlife Conservation. It may have been lost for 50 years, but it is possible that *Mystacina robusta* will be found again. There is no such hope for the bush wren and the South Island snipe.

Some good came out of the tragedy of Taukihepa. It opened the eyes of many in government and wildlife management to the threat that rats pose to native wildlife—these pests could, indeed, wipe out whole species. The success of the South Island saddleback rescue saved this special bird, and now around 2000 birds survive on small island sanctuaries in the south, and at the Orokonui Ecosanctuary near Dunedin. All of these are descended from the 36 birds saved from Taukihepa in 1964.

Merton and Bell and their co-workers would go on refining their capture and translocation techniques and use them to save other species such as the black robin of the Chatham Islands, which was brought back from the brink of extinction (see chapter 9). New Zealand would go on to lead the world in endangered species conservation and recovery—but it was a hard way to learn the lesson.

Tāiko (*Pterodroma magenta*)

9. No island refuge: The strange—and vanished—species of the Chatham Islands

Charles Fleming held on tight to a crack in the rocks and tried not to think about what would happen if his fingers slipped. The climb up the cliffs of Little Mangere Island was more arduous than he had imagined, even after he'd looked up at the rocky walls he would have to scale from sea level. It was no wonder that the rare birds of this island had been left alone for so long—you'd have to be crazy to climb up here. And yet here he was.

Little Mangere was little more than a rock in the middle of a tumultuous ocean. Fleming and his companions had made the journey across the turbulent waters from Chatham Island to Mangere and Little Mangere, tucked near to the side of Pitt Island, in a fishing boat. It had taken them four attempts to get ashore, and now they had to climb up this cliff.

They weren't sure what they would find once they had scaled the near-vertical cliffs to the small patch of wind-blasted bush on the summit—only a few hardy locals and the odd rare-bird collector had visited the island for 30 or more years. But there were three birds they really hoped to see. Earlier visitors had written of a tiny black robin, a funny little bird with entirely dark feathers; and of a large and noisy parakeet, different to the red-crowned variety found on the other islands. There had also been reports of the Chatham Islands bellbird being seen back in 1906, after it had vanished from the other islands. But none of these three species had been seen for a long time. They were assumed to be extinct.

After an hour of climbing, past stinky and noisy colonies of shags and gulls precariously perched on the rockface, Fleming came to some small patches of scrub, hanging on to the cliff as tightly as he was. And suddenly there they were, bobbing around in the branches: a pair of tiny black robins. Fleming felt his excitement rising.

Once he and his companions had reached the forest on the island summit and started to walk through the undergrowth, stumbling knee-deep into the burrows of petrels dug into the sandy soil, robins followed them. Sometimes it was just one bird and sometimes a pair, curious and apparently fearless, coming to within an arm's reach. As the walkers left each robin's territory, a new pair would carry on the supervision, darting around the men's heads and swooping down to snatch up insects disturbed by their footsteps. There seemed to be at least twenty pairs of this bird which had been lost to the world for three decades.

'Just as well this island is so hard to get to,' Fleming thought, pulling his leg from yet another muttonbird burrow. 'Surely the birds will be safe here.'

Out of sight, and to many New Zealanders largely out of mind, the Chatham Islands lie nearly 900 kilometres east of Christchurch. Legally and geologically part of New Zealand, the islands nevertheless remain separate and apart, surrounded by the ferocious Southern Ocean. They're actually so far to the east of New Zealand that they're in a different time zone—clocks in the Chatham Islands run 45 minutes ahead of 'mainland' time.

There is one large main island, the 920 square kilometre Chatham Island, called Rēkohu by its original Moriori inhabitants or Wharekauri by Māori; one much smaller one, Pitt Island (7.7 square kilometres), known as Rangihaute to Moriori and Rangiauria to Māori; and a scattering of even tinier ones, including Mangere Island and Little Mangere Island (Tapuaenuku), South East Island (Rangatira), the Pyramid (Tarakoikoia) and the Forty Fours (Motuhara), which are little more than a cluster of wave-battered rocks lying on the 44 degrees South line of latitude. The islands are low-lying and windswept, but were once home to a unique collection of birds and plants, some found nowhere else in the world. As with the main islands of Aotearoa, humans and the pest animals they brought with them spelled doom for many of these previously isolated wonders.

When it comes to extinction statistics, the Chatham Islands are, unfortunately, world-leading. While around 24 per cent of mainland New Zealand species have become extinct since human settlement, on the Chathams the figure is greater than 50 per cent. From giant swans and chubby flightless ducks to secretive rails and tiny snipe, more than twenty different types of bird that once lived on these islands have disappeared from the world entirely in just a few hundred years. Of these, at least ten different types (both species and subspecies), including five found solely on the Chathams, were wiped out following Polynesian settlement of the islands; a further eight have been exterminated since

Europeans came, including four found nowhere else. Many of those that remain are endangered. It is a sad record to hold.

Many of the bird species that persist on the Chathams resemble their mainland counterparts, but with subtle differences. Some are a different size or a slightly different colour. Surviving members of this unique treasure box of birds include the Chatham Island pigeon, or parea, which is both plumper and paler than the kererū; the Chatham Island oystercatcher, smaller and stockier than the variable oystercatcher; and the Chatham Island tūī, which is larger than the mainland tūī, with a longer throat tuft and a different song.

It is now believed that the Chatham Islands were settled by Polynesian people who crossed over from the South Island during the 1400s. Once on the islands and isolated from Māori on the mainland, the people developed a different language and culture, including a social code which banned killing in warfare. These Moriori people, as they came to be known, lived largely off the bountiful seafoods of the islands and the kernels of the berry of the kōpī or karaka tree, which they brought with them and cultivated in special forests.

They preyed on the islands' birds, too, and the first losses of species came during this period. Te Whanga Lagoon on Rēkohu was thickly populated with tasty waterfowl, including a sturdy and possibly flightless duck (*Pachyanas chathamica*); a native black swan, similar to but larger than the Australian variety introduced in the 1860s (named poūwa, or *Cygnus sumnerensis* by scientists); a coot (*Fulica chathamensis*), which probably looked a lot like a black pūkeko; and a shag-like bird called a merganser (*Merganser australis*), which persisted on the subantarctic islands to the south of New Zealand into the early twentieth century. There was also a large bird that looked similar to a weka, called Hawkins' rail (*Diaphorapteryx hawkinsi* or mehonui), and a snipe (Forbes' snipe or *Coenocorypha chathamica*), both of which are known only from fossil bones but which may have survived into the 1800s; an island variety of kākā (*Nestor chathamensis*); and even a raven (*Corvus moriorum*), which hunted seabirds and scavenged among seal colonies. While the Moriori people ate adult birds, the kiore they brought with them preyed upon eggs and chicks.

Some people used to think that the skies of the Chathams were patrolled by quite a different predator—a bird of prey. Scotsman Henry Ogg Forbes, director of the Canterbury Museum, visited the island in 1892 and collected large numbers of fossil bones, which he took back to England with him. In 1961, a New Zealand researcher called Elliott Dawson, who had been working through Forbes's many finds at the British Museum, suggested that they included the bones of a sea eagle 'quite unlike . . . the White-bellied Sea Eagle (*H. leucogaster*) ranging into

Australia and western Polynesia'. In fact, the bones closely resembled those of the bald eagle found in northwestern Canada and Alaska.

And that's because they probably were. Forbes also collected bones from British Columbia, Canada, and many of his finds were labelled only after his death. No further fossilised eagle bones have ever been found on the Chathams, and it now seems likely that these intriguing remains were Canadian bald eagle bones that had been incorrectly labelled.

The special species of the Chatham Islands were further devastated by Europeans after the first sailors arrived in 1791. The British survey ship HMS *Chatham* came across the islands by accident, when it got blown off course. Captained by William Broughton, the *Chatham* had been heading for Tahiti when it happened upon the northwestern corner of what was subsequently named Chatham Island in the middle of a wild, stormy ocean.

Following this visit, it took a few years for word to get out about the abundance of seals and sea lions there—but once it did, the fate of the islands and their inhabitants, both animal and human, was confirmed. The first sealing gangs came to the island in the early 1800s and the seals were hunted out within a few decades. Attention then turned to the right whales and sperm whales found in the surrounding waters. Shore-based whalers and sailors who ran away from whaling ships were the first Europeans to live on the islands.

The new European arrivals brought diseases with them, which killed many Moriori. Then, in 1835, a group of Ngāti Tama and Ngāti Mutunga Māori, who had migrated from Taranaki to Te Whanganui-a-Tara (Wellington) during the Musket Wars of the 1820s, decided to move on and colonise the Chatham Islands. They used their guns and force to overpower the remaining Moriori, killing many and taking others as slaves, and claiming the islands for themselves.

In 1840, the New Zealand Company (a British company that established European settlements here and encouraged immigration) purchased the islands from the new Māori occupiers, with the intention of establishing a permanent settlement there. This deal was soon overturned and the islands were officially made part of the new British colony of New Zealand. Some missionaries and a few hardy European farmers came to live on the islands, and the destruction of their natural habitat to create farmland and provide timber accelerated.

All of this change spelled disaster not only for the Moriori people—the last full-blooded Moriori, Tommy Solomon, died in 1933—but also for the unique and precious species that also made their homes there. Rats, cats, pigs and possums ran riot on the isolated islands, rapidly erasing many rare and unusual birds found nowhere else in the world.

The first naturalists to visit the islands found birds similar to those on the mainland but with significant differences. German naturalist Ernst Dieffenbach, who visited the islands in the winter of 1840 with the New Zealand Company expedition, wrote the first description of the islands to be published in Europe. In the journal of the Royal Geographical Society in 1841, Dieffenbach wrote that while the seals, which were once plentiful on the island, had been all but wiped out,

> [t]he birds are more numerous. Vast flocks of the common dark grey duck, snipes, plovers, curlews and redbills inhabit the lakes and sea-shores, and a sand-lark which builds its nest on the ground, abounds in the rushes of phormium [flax] and fern . . . The forest is enlivened by numerous tuis or mocking-birds; a little green perroquet [parakeet] flocks in hundreds to the potato-fields, and proves a great nuisance to the farmer by picking up the seed as soon as it is sown. This bird is generally a little larger than the New Zealand parroquet, and is perhaps a different species.

Dieffenbach also observed 'mako-mako' (bellbirds), 'the finest songster in New Zealand'—again larger than the mainland type; kererū (wood pigeons) and 'three or four small, insectivorous birds'.

> A new kind of rail was formerly very common; but, since cats and dogs have been introduced, it has become very scarce. The natives call this bird meriki, and catch it with nooses. I often heard its short, shrill voice in the bush, and, after much trouble, obtained a living specimen.

It is possible that this bird, from which the species known as Dieffenbach's rail (*Gallirallus dieffenbachia*) was described, was one of the last of its kind. This was almost certainly the last time one was seen. The flightless rail, which was only about the size of a tūī, had once been found on the Chatham, Pitt and Mangere islands, but was now lost forever. Although it had survived human occupation of the islands for around 400 years, the arrival of European pests finished it off soon after Dieffenbach took his specimen.

Two other species of rail which were also found solely on the Chathams suffered a similar fate. One was Hawkins' rail, described earlier, and the other was Hutton's rail (mātirakahu or *Cabalus modestus*), which was much smaller, flightless and nocturnal. It was first scientifically described from specimens collected by Henry Hammersley Travers in 1871. Captain Frederick Hutton of the Colonial Museum in Wellington noted that 'this curious bird' with its curved bill was found solely on Mangere Island. Exterminated from the other islands in the group by rats and other pests brought to the Chathams by humans, Hutton's rail was eliminated from its final refuge in the 1890s after cats found their way ashore there.

As Dieffenbach had noted, the Chathams were also home to a local form of the bellbird (*Anthornis melanocephala*), and a small, shy type of fernbird (*Bowdleria rufescens*). When Travers visited the islands, he found the bellbird on the main Chatham Island, Mangere Island (in the greatest numbers) and Pitt Island. He described its call as being 'much richer and fuller' than that of its mainland cousin. By the turn of the century, however, the bellbird could be found only on Little Mangere, and by the late 1930s it couldn't be found at all. It was a similar story for the fernbird. Travers described it as being 'not uncommon' on Mangere Island but 'difficult to secure' because of 'its peculiar habit of hopping rapidly from one point of concealment to another'. Just two decades later, Henry Ogg Forbes considered it to be extinct, another victim of the wild cats on the island.

The birds of the Chathams also shared another threat with their rare and unusual counterparts on the mainland. They were sought after by collectors keen to gain specimens for private and museum collections. While the obtaining of specimens for British collector Walter Rothschild (see chapter 4) in the 1890s enabled 'a good deal more information' to be obtained about the unique birds of the Chathams, it also drove a few of them to extinction. The last fernbird known to the world was taken from Mangere by Rothschild's collector in the 1890s. Rothschild's museum at Tring Park, northeast of London, also contained 'a large series' of examples of the Chatham Island bellbird. Unfortunately, the last few were seen on Little Mangere in 1906, and now the species has disappeared from the world forever.

But the story of the unique species of the Chathams is not all doom and gloom. Alongside the losses stand two amazing stories of survival—one of a species being brought back literally from the brink of extinction, and the other of a seabird being rediscovered after being 'missing, presumed dead' for more than a century.

The black robin of the Chatham Islands, or kakaruia (*Petroica traversi*), is a tiny, rounded, all-black bird with beady eyes which was originally found on the five main islands in the group. The bird was first described by Travers following his 1871 visit. Travers found the bird solely on Mangere Island, but in 1892 Forbes also obtained specimens from Little Mangere.

The bird was then not seen for more than 40 years, until a university student called Charles Fleming and his mate Graham Turbott made an expedition to the Chathams in late 1937–early 1938, along with a schoolteacher from Kaingaroa on Chatham Island, Allan Wotherspoon. The three men scaled the rocky sides of Little Mangere and found between 20 and 35 pairs of robins there. (They also rediscovered Forbes' parakeet, a special type of kākāriki, which was also thought to be extinct.) In an article about the expedition, Fleming wrote of the challenges in accessing the steep-sided, rocky island:

> Future investigators who might wish to spend time on the island . . . should be prepared to erect rope ladders and carry all drinking water. Transporting gear up the four hundred foot [122 metre] cliff would be possible, but difficult.

He was also the first person to mention the possibility of translocating some of the robins to another island to ensure their survival.

After Fleming and Turbott's visit, the birds of Little Mangere were left to their own devices again for many years. In the 1950s and 1960s, the Wildlife Service then turned its attention to the Chathams and their dwindling cargo of rare species, and took over Mangere and Rangatira islands as nature reserves. The wild cats had been exterminated by farm workers visiting the islands to shear the last few wild and woolly sheep living there, and the islands were both completely pest-free. In the 1960s the last of the livestock was removed, and work began to reforest the island and re-establish populations of Chatham petrel (ranguru), shore plover (tūturuatu), Chatham Island snipe, Chatham Island oystercatcher (tōrea) and parea (Chatham Island pigeon) on these islands. Forbes' parakeet and the little black robin remained isolated on the rocky stack of Little Mangere.

By the early 1970s, studies showed that only around twenty black robins survived. Their numbers continued to fall, and when the population reached just seven birds in 1976, a decision was made to transfer the birds to the larger Mangere Island, where 120,000 trees had been planted to create shelter. Even then the robins continued to struggle, and in 1980 there were just five of these tiny birds left anywhere on the planet, of which only two were female.

Desperate times called for desperate measures. A Wildlife Service team including Don Merton and Brian Bell (see chapter 8) pioneered new techniques of 'cross-fostering', which involved taking eggs laid by the female robins and moving them to the nests of Chatham Island warblers and tits. The 'foster parents' would then incubate and hatch the robin chicks while the robins were encouraged by the 'loss' of their eggs to lay more.

Not all the eggs and chicks survived the early experiments, but that first season four chicks made it to fledging (developing wing feathers large enough to enable them to fly)—nearly doubling the black robin's global population in one summer! Three of the chicks were the offspring of one female, nicknamed 'Old Blue' because of the colour of her identifying leg band. Despite being at least nine years old, Old Blue had laid three clutches of eggs in one season—and almost singlehandedly saved her species from extinction.

And so began the black robin's long, slow climb back from the brink of oblivion. Intensive management throughout the 1980s saw the population climb, slowly but surely. Some birds were moved to Rangatira, including Old Blue. After making her significant contribution to saving her species, she was last seen alive in late 1983.

Today there are around 240 black robins on Mangere and Rangatira islands. Conservationists hope that one day they will be able to return to larger islands in the Chathams group and even to Little Mangere. This little island is still the stronghold of Forbes' parakeet, which has also recolonised nearby Mangere.

Don Merton might not have been able to save the South Island snipe and the bush wren, but there is no doubt that his actions and those of his team saved the black robin from disappearing from the world forever. Merton passed away in 2011, but his legacy lives on in the black robin, the kākāpō, the tīeke and the other birds which he helped to save.

Just as the work to save the black robin was beginning in the 1970s, another scientist was also hard at work on the Chathams. He wasn't trying to save a species, however; he was trying to prove that one which the world believed extinct still existed, hidden from human sight by its nocturnal and seagoing habits.

The first European record of the bird now known as the Chatham Island tāiko was made in July 1867, when scientists aboard the Italian research vessel *Magenta* captured and killed one in the southern Pacific Ocean, about 800 kilometres east of the islands. They also saw the same type of birds even further afield, south of Easter Island and off the coast of Chile. The dead specimen was taken back to the University of Turin, in northern Italy, and declared a new species: *Pterodroma magentae*, the magenta petrel. However, no one ever saw the handsome grey bird with a white underside again.

When Charles Fleming visited the Chatham Islands in 1938, he recorded stories of a breeding colony of large petrels in the southwestern corner of the main island. These birds had been taken as 'muttonbirds' by the local people until the early 1900s, but 'of late years visits to the breeding grounds have ceased for, with the increase in vermin, birds have become fewer, and some of the colonies are now verging on pasture land'. The local people called these birds tāiko, or tchaik in the Moriori language. Like the other types of petrel which bred on the island, we now know that they spent their winters at sea, ranging far across the Pacific, before returning to the island in summer to breed in burrows under forest cover.

In 1952, an enthusiastic schoolboy called David Crockett spent time helping Canterbury Museum's fossil bone specialist Ron Scarlett, working with specimens from the Chatham Islands. Crockett found some unidentified bones from midden sites which did not relate to any species of petrel known to live on the Chathams. He spoke to Fleming, who suggested that he get in touch with Chatham Islands farmer Henry Blyth, whose land was home to the mysterious petrel colony. Blyth wrote back to Crockett to say that while he knew of two colonies on his land which had disappeared, he believed there was still a third one, in an inaccessible part of his farm known as Tuku Gully.

Seventeen years later, in 1969, Crockett made it out to the islands to look for himself. Now a school science advisor, he was fascinated by the idea of tracking down this elusive petrel. Blyth was no longer alive, but Crockett spoke to other Chatham Islanders and heard their stories of harvesting the mysterious petrel. They believed that the birds were long gone and told him he was hunting a taipō—something ghostly or supernatural—rather than a tāiko.

Crockett went back again at the end of 1972, and in January 1973 he caught his first glimpse of the mystery bird. One night at his base camp, four dark birds with white bellies were attracted to the beam of a floodlight set up above the gully where Crockett believed the birds were nesting. They swooped silently in the beam for about twenty minutes, then vanished.

The sighting excited Crockett's interest still further. But it took another five years and three more expeditions to the Chathams before he and his companions were finally able to capture and identify a live bird. On New Year's Day 1978, they finally captured, photographed, measured and released two specimens. The tāiko was no ghost.

Today, after more research and management of the tāiko's breeding habitat, its population is slowly growing. Tuku Gully is now a reserve, after landowners Manuel and Evelyn Tuanui gifted more than 1000 hectares of bushland to the government. The Chatham Islands Tāiko Trust, in partnership with the Department of Conservation, works on predator control, killing mostly feral cats. A second breeding area, at Sweetwater, is protected under covenant by Manuel and Evelyn's son Bruce Tuanui. It has been fully predator-fenced and birds have been transferred there to breed in a safe environment. Tāiko breeding is closely monitored, using infrared cameras and electronic tags, and additional food is supplied to chicks if required.

Tāiko lay just a single egg each season, so any population growth is painfully slow, but in the breeding season of 2018–19 a record number of 27 chicks were fledged from a known 35 breeding pairs. It is now believed that there are around 200 birds alive.

The tāiko and the black robin are just two of the many special species of the Chatham Islands which still face challenges to their survival. Conservation efforts are also focused on saving the islands' rich biodiversity of native plants, lizards, freshwater fish and invertebrates, including endangered species of spiders, stick insects and beetles.

Conservation work on the islands continues, including monitoring and eliminating pests and replanting forest cover on Rangatira and Mangere islands and at reserves on the main islands. None of these species is yet out of the woods, but at least they are still here, and being taken care of.

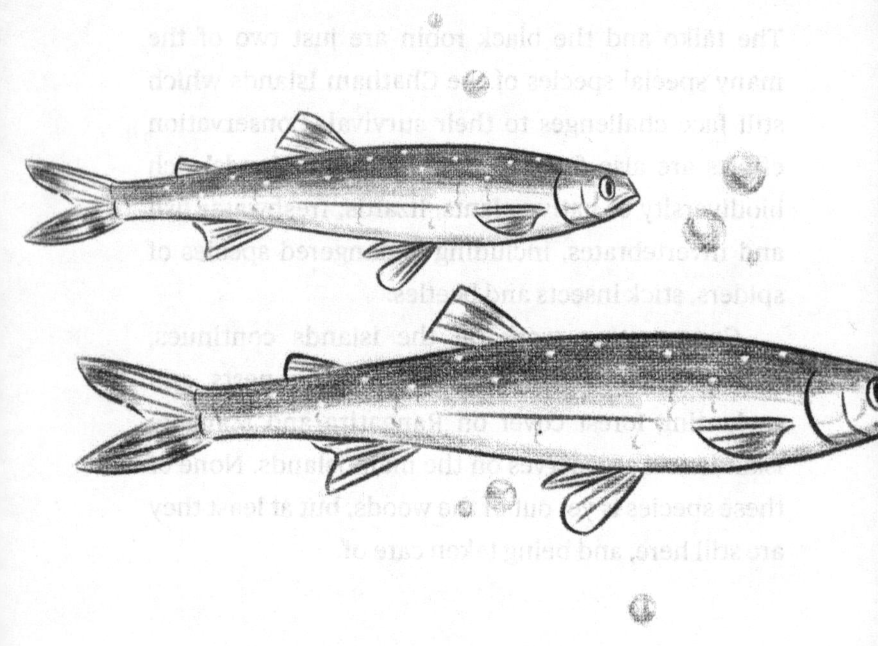

Grayling (*Prototroctes oxyrhynchus*)

10. Plenty more fish in the sea: Grayling

Maggie stood with her toes in the edge of the river. She had been digging the garden for hours and needed a rest. The water looked cool and inviting. But as she stood on the bank, looking across the broad expanse of water to the thick bush on the other side, she could hear one of the children crying. She'd better go get supper on the table before the men returned from the bush.

She noticed a strange movement in the water, like it was coming to the boil. In the early evening light, the surface of the water began to sparkle and dance where the river widened and slowed as it turned a long corner. At first Maggie couldn't make sense of what she was seeing—what on earth could make the water move like that? Then she realised the patterns and splashing were being made by fish—hundreds, maybe thousands of them—congregating in the river pool. The last of the sunlight caught on their scales just under the dark skin of the water, and on their metallic tails and fins—sometimes silver, sometimes gold—as they broke the surface and threw droplets of water in the air. Maggie had never seen anything like it.

The shoal moved as one, ebbing and flowing around the pool, coming almost to her feet. Now she could see the fish properly—larger than her hand, the size of a good trout. The fish seemed unaware of her presence, completely consumed by their watery dance.

Maggie heard another cry from back at the house. Reluctantly, she turned away from the spectacle in the river. It'd been a bright spot in an otherwise humdrum day. She couldn't wait to tell George about it.

George, of course, would ask her why she hadn't tried to catch them. She didn't think he'd understand if she told him the fish were too beautiful to kill.

Most New Zealanders can name quite a few of our extinct native birds (and hopefully you can recognise even more of them after reading this book). But few Kiwis now remember the story of the grayling, the only freshwater fish in this country to suffer the same fate. It is the only species of native fish to currently enjoy legal protection, even though no one has seen one for nearly a hundred years. Learning from the fate of the grayling may be critical for our surviving species of native fish, many of which are considered at risk of extinction.

The grayling (*Prototroctes oxyrhynchus*) was a species well known to Māori, who knew them as upokororo, kanaekura, tīrango or paneroro. Young fish were also called nehe. Grayling made good eating, and their abundance made them easy to catch. They were decent-sized fish, growing up to 40 centimetres long—about the legal size of a rainbow trout, and a similar shape. The species name means 'pointed snout', and remaining museum specimens show it to have a rounded belly and a pointy nose.

The grayling was silver in colour, but contemporary reports suggest that it became darker with age, and that larger specimens were more golden red on top. It was related to the Australian grayling, which also suffered a population decline in the nineteenth century, although the New Zealand species was bigger. The Australian fish shares a bizarre quality with its extinct Kiwi cousin: when caught, they smell like cucumber. Unlike other native fish, which are omnivorous, grayling ate only plant matter, grazing on algae and other growths on riverbed rocks. One writer said 'their presence could be detected by their teeth-marks on the stones'.

The first Māori method of fishing for grayling recorded by a European observer seems both labour-intensive and unlikely. Missionary the Reverend Richard Taylor wrote in 1855: 'The *upokororo* is a fish about eight inches [20 centimetres] long, with scales; it is caught in the autumn; it bites at the hair of the legs, and is thus caught by the natives going into the water.' No further details were supplied by Taylor as to how the fish were actually captured, and if they remained attached to the fishermen's leg hairs!

In the late 1920s, ethnologist Elsdon Best recorded some other traditional methods used to catch upokororo. In the mighty Waikato River, they were taken in nets 'many fathoms in length' (a fathom is 1.8 metres). On the East Cape, fish were caught in hoop nets made with supplejack frames, set across narrow sections of the river. Fish were then chased downstream into the net. Sometimes trenches were dug along sandbanks and the fish driven into them. Artificial weirs (low dams) and stone walls were sometimes built to trap the fish behind, then the river water diverted so that the fish, high and dry, could be collected.

Pākehā fishers, who named the grayling after an unrelated freshwater fish found throughout Europe, preferred to catch them on a line, although they were not a true game fish like the trout and salmon the new immigrants were used to. Newspaper articles describe grayling being caught with worms or on artificial flies, giving 'splendid sport when in the humor to feed'. In 1870 geologist Julius von Haast noted that a correspondent had told him of 'Our most skilful brother of the rod, Mr. J. H. Smith . . . as shown by his diary, caught last year 1,152 of these fish in fifty-eight fishing-days, taking in one day ninety-three'.

Although the fish were always taken in freshwater, early observers assumed—correctly, as it turned out—that upokororo spent part of their life cycle at sea, like other native fish species such as smelt and galaxiids such as īnanga. Scientists now believe that the grayling spent much of the year in rivers and streams, feeding and spawning, then in spring the young fish would be washed out to sea for several months before returning to freshwater.

Up until the late 1800s, grayling were abundant. Like the passenger pigeon of North America, which once swarmed in its millions before it was hunted to extinction, upokororo were once seen in huge shoals in rivers and streams throughout the country. There were so many of them that their dead bodies were sometimes used as fertiliser on market gardens.

The appearance of a shoal of grayling often resulted in a few column inches in the local paper, such as this report from the *Wairarapa Daily Times* in June 1885:

> A shoal of upokororo fish was seen in the Waipawa [Waipoua] river on Thursday afternoon, in the deep hole by the southern end of the

> railway bridge. This is not the season in which these fish are usually expected, and about fifty natives at once came over from the pah [pā] to endeavour to catch them.

In May the previous year, a shoal of around a thousand fish had been seen in the same river, 'varying in size from 4oz [100 grams] up to 21bs [9.5 kilograms]'.

However, just a few years later the fish seemed to be vanishing. The big shoals which were once common were no longer being reported, and single catches were newsworthy. When a grayling was hooked in the upper Ōtaki River, running out of the Tararua Ranges, in January 1906, the catch was reported around the country, with the fish being described as 'once plentiful . . . but now almost extinct'. However, 'Not knowing what a valuable catch they had taken, the party cooked and ate this survivor of old time'. In 1921 another fish, weighing about half a kilogram, was caught by a fly-fisher in the Waiwhakaiho River, just north of New Plymouth (Ngāmotu).

In 1923, a party including doctor and anthropologist Peter Buck and Elsdon Best spent a month on the East Cape north of Gisborne gathering information on Māori culture and beliefs, including making phonographic recordings of traditional songs. Films and photographs were also taken of traditional fishing methods, and a surprising discovery was made:

> After getting some instructions from the Natives in the art of netting, the members of the party set up some traps in the rapids of the Waiapu River. The object was to catch kokopu, but instead of capturing that fish they were surprised to find that in their net were no fewer than 42 specimens of a fish now rare in this part of New Zealand, the upokororo . . . The catch of the now rare upokororo was a great surprise to the Natives, and the news spread along the coast from village to village.

This is now accepted as the last time upokororo were seen anywhere in New Zealand. The once abundant shoals of silvery grayling were gone.

Bizarrely, the grayling was legally protected only decades after the last confirmed catch. The new Freshwater Fisheries Regulations that came into law in 1951 included a special section on 'Indigenous Fish', which stated that 'no person shall intentionally fish for, take, or kill grayling or fish of the genus *Prototroctes*'. If a grayling was caught by accident, it had to be reported. The law also rather vaguely stated that 'No person shall in any water intentionally kill or destroy small indigenous fish other than elvers' (immature eels)—then added a note to say that taking whitebait (the young of several different species of native fish) for human consumption was allowed. The grayling remains fully protected by law even today; the whitebait species still are not.

No one knows why the grayling died out. Its decline was relatively rapid but also widespread—it disappeared from most New Zealand rivers at around the same time, over a period of little more than a decade. Contemporary observers blamed the introduction of brown and rainbow trout to its river environments from the late 1860s, but often the fish seemed to disappear before trout were strongly established, and also from rivers where there were no trout. Fishing for sport must have had some impact on population numbers, but it was not excessive. There was also the impact of changes to the greylings habitat due to forests being cleared and farmland being established.

In 2019, environmental scientist Finnbar Lee of Auckland University conducted a study of the grayling's decline. He concluded that it was a victim of 'multiple interacting factors'—many different things at once—including overfishing, being preyed on by trout, and a decline in river quality. Because grayling did not necessarily return to the same river each season, they died out over a wider area, more quickly, as more rivers became unhealthy for them.

The downfall of the grayling is significant for the ongoing survival of our remaining freshwater fish species. New Zealand has 56 species of native freshwater fish, many of which are found only here, and nearly three-quarters are considered threatened or at risk of extinction. Fortunately, in October 2019, the government passed a bill amending current conservation legislation to better protect native fish species. The new laws will allow better management of the main threats to indigenous freshwater fish, such as barriers to fish passage (like drains and culverts) and loss of spawning sites, as well as controlling activities such as drain clearing which can kill large numbers of eels and other fish.

Among the most critically at risk are the galaxiid species, including giant kōkopu, shortjaw kōkopu, īnanga and kōaro, the young of which are known collectively as whitebait. While whitebaiting is currently restricted to certain seasons and times of day, management of the fishery is currently being reviewed, and it seems likely that changes will be made to protect declining populations. Continuing to harvest and eat the young of these rare species may no longer be sustainable.

You wouldn't eat a baby kiwi—should we really go on eating whitebait?

Part 2
Lost and found

Takahē (*Porphyrio hochstetteri*)

11. The kings of hide and seek: Takahē and moho

It was nearly dark, and the sealers were tired. The days were short and cold at the bottom of the Middle Island, and a light dusting of snow lay on the bush around them, right down to the waterline. The seals were becoming fewer and farther between, and this day they had managed to kill only ten. To add to their misery, they'd also failed to shoot any small birds to add to the pot.

'Fish for supper again tonight,' one of the men, Hobbs, sniffed as the party worked their way back round the foreshore, feet crunching on the icy gravel. 'Wouldn't half mind a roasted chicken.'

'What about that, then?' His companion, Burton, pointed to a footprint in the snow at the edge of the treeline. 'That looks like a mighty big chicken.'

Hobbs paused in his march and crouched down by the print, the impression of three long toes pointing forward and a shorter one pointing back. 'Go on, Tom—find the bird,' he said to his dog, which had started sniffing around in the underbrush. It wagged its tail obediently and bounded into the trees.

After a minute or so the men heard the dog bark, then a series of tremendous shrieks tore through the still air. Hobbs wondered what on Earth it had found.

Hobbs and Burton pushed their way through the dense bush in the direction of the cries, and came across Tom bailing up a large and spectacular bird. It stood nearly two feet high, with handsome dark-blue and green feathers and a bright-red beak and feet. The bird was screaming and snapping at the dog dancing around it.

Hobbs reached for his gun, but Burton put a hand on his arm.

'Let's try to catch it,' he said. 'Might fetch a decent price if it's something unusual.'

'You reckon?' Hobbs sighed, but he sensed his friend was right. It was like no other bird they had seen before in this wilderness. 'Tell you what—let's try to take it back to the boat alive. If it causes us any trouble . . .' He drew a finger across his throat. 'I reckon it'd be right tasty. Plenty of meat on that thing.'

Saliva rushed into Burton's hungry mouth. His mate had a point.

When a party of sealers caught a large, unusual and brightly coloured bird in Dusky Sound (Tamatea) in 1848, it was the first time Europeans had laid eyes on what we now know as the takahē. One of the key conservation success stories used to showcase New Zealand's ability to bring species back from the brink, takahē can now been seen on sanctuary islands just an hour by boat from Auckland city. But for many years they stayed hidden from human sight in the deep, deep south, in valleys where few—if any—humans have ever walked.

Young Englishman Walter Mantell didn't even know he'd found the skull and other bones of a new species when he dispatched a box of moa and other bones to his father in England in 1847. Mantell had uncovered the bones in a Māori midden site on the Taranaki coast, and thought that his father, Gideon, might find them interesting. Gideon (who we met back in chapter 2) was a well-known doctor with an interest in fossils. He had wanted his son to follow in his medical footsteps, but Walter had emigrated to try his luck in the new colony of New Zealand. He arrived shortly before the signing of the Treaty of Waitangi in February 1840. Walter Mantell worked in various lowly positions in government, which gave him plenty of time to pursue his passion for natural history.

Upon receiving the bones, Gideon Mantell recognised that some of them were unusual. He showed the small bones to anatomist Richard Owen —he of the moa 'discovery' (see chapter 2)—and the respected expert declared them to be the bones of a new species, which he named *Notornis* (meaning 'southern bird') *mantelli* (after Walter). It was considered to be yet another of the mysterious extinct species of the new colony, already lost to the world.

That changed two years later, when young Walter Mantell was again a key player in the story of the *Notornis*. By this time he had been appointed 'commissioner for extinguishing native titles, Middle Island', which meant that he travelled the South Island purchasing Māori land for the Crown and convincing Ngāi Tahu, who lived there, to move on to 'native reserves' set aside for them. This travelling allowed him to continue collecting natural-history specimens, and in late 1849 he was offered some bird skins collected by a sealer. Among them was the skin of what he believed must be a *Notornis*—which had until recently been very much alive. A sealing party had caught it in Dusky Sound, then after keeping it captive for a couple of days had eaten it, declaring it 'very good . . . as it was the first of its kind, all hands had a taste for curiosity'. Mantell shipped the skin back to his father, who figured out that the skin came from the same type of bird as the earlier bones. The discovery was greeted with much excitement. Here was evidence that a bird which had been believed to be extinct still walked the Earth. It was as if someone had found a live dodo strolling around.

However, it appeared that *Notornis* were still few and far between. It was another couple of years before a second skin was procured—a bird shot by a party of whalers in Fiordland—and a third was not caught until 1879, near Lake Te Anau. One further bird was captured by a dog in 1898, also near Te Anau, but after that the number of possible sightings dwindled. Hunters, trampers and back-country workers made occasional claims to have seen the birds, but the official government line was that 'in the interests of conservation' their location should be kept a secret, so that they were not hunted or otherwise 'interfered with'.

Perhaps these last four specimens had been outliers, the last few birds of a species doomed to extinction. It would indeed be nearly half a century before the bird was seen again.

There was once a North Island takahē, too, which the Māori knew as moho, meaning lonely. Fossil records show that the birds lived throughout the island, in forests but also in swampy areas and sand dunes. While many people think of takahē as an alpine bird, they were once happy to live right down at sea level. Strictly speaking it was the bones of a moho, not a takahē, that Walter Mantell found at the Waingongoro River mouth in South Taranaki. It is this bird which retains the Latin name *mantelli* (though it is now *Porphyrio mantelli*), while the South Island species is named *Porphyrio hochstetteri*, after the Austrian geologist Ferdinand von Hochstetter, who was the first to analyse and describe many of New Zealand's underlying rocks.

Subsequent research has proved that the two birds are distinct species. It seems likely that the birds shared the same colouring, with carmine-red beaks and legs, a dark, glossy blue head, neck and chest, and an olive-green back—much more vibrant than their relatives the pūkeko, though with the same flash of white under the tail. The moho was larger than the takahē, with longer legs and a more 'slender' look than its plump relatives, which grow to around 50 centimetres long and up to 3.5 kilograms, the size of a large chicken. Like the southern species, the northern bird probably ate native grasses, stripping the base of the leaves to get to the new growth. Studies have shown that in an alpine tussock environment, takahē need to eat almost constantly to get enough nourishment from the snow grass.

Moho bones have been found in midden sites, suggesting that they were eaten by Māori and possibly hunted to extinction by them in the early settlement period. There is only one report of a moho being seen by European eyes. In 1959, W. J. Phillips wrote in the journal of the New Zealand Ornithological Society of 'an important and apparently factual occurrence of a strange bird taken in the Ruahines last century'. The story went that in the 1890s, 'an actual specimen of *Notornis* had been secured in the Tararuas and taken to the home of Mr. Roderick McDonald of Horowhenua'. Phillips searched out Hector McDonald, Roderick's son, in early 1959, and asked him about the tale.

> As a young man Hector McDonald was present when the bird arrived at the McDonald homestead. There was great excitement at the time, particularly among the older Maori population who identified the bird as the rarely seen Mohoau [moho]. This was in the autumn of the year 1894. The skin and feathers were kept at the homestead for many years.

The moho had apparently been 'secured' by surveyor Morgan Carkeek, who had been 'working in the north Ruahine Ranges, and the bird (or its skin) was brought through with other equipment on pack horses'.

Was it indeed a moho? Was it possible that these mysterious birds could have lingered on in the mountains of the North Island as well? Without hard evidence, we will probably never know.

The continued existence of the southern takahē was confirmed in spectacular fashion in 1948, however. Invercargill doctor Geoffrey Orbell had become interested in the legend of the big blue birds when he was a child, after his mother showed him a photograph of a stuffed takahē in the Otago Museum (the 1898 Te Anau bird). A keen tramper and back-country explorer, Orbell's obsession with the takahē grew with him into adulthood, and in April 1948 he and two friends, Rex Watson and Neil McCrostie, went into the Murchison Mountains west of Lake Te Anau to look for the mysterious birds.

On that first expedition they didn't see any live takahē, but they found footprints and droppings that they believed had been made by the birds. Orbell was encouraged, and following a cold and snowy winter the party, now joined by Watson's girlfriend Joan Telfer, went back into the mountains in mid-November.

This time their efforts were rewarded. After an arduous climb into the mountains from Lake Te Anau, they reached a tussock flat. Suddenly Orbell dropped to the ground and hissed, 'It's either a damned big swampy [pūkeko], or it's it.'

It was indeed a takahē, 'strutting about on a patch of swampy ground', Telfer recalled. 'The bright crayfish beak and dark head were unmistakable.'

The bird seemed completely unconcerned by the presence of humans, and when a second one joined it they were easily netted. Tethered to stakes, they 'quickly settled down' for the rediscoverers to watch and photograph. 'When we completed our observations and the birds were released, we returned in what *Time* magazine so aptly described as "a state of ornithological ecstasy",' Telfer wrote in 1949.

The cat—or, rather, the bird—was out of the bag. Once again the takahē had risen from the dead.

Orbell's fantastic find created headlines around the world. It became apparent that there were between 30 and 40 living takahē, spread across several valleys. A 650 square kilometre area of the Murchison Mountains around 'Takahē Valley' where the birds had been found was declared a 'Special Area' by the government, and could only be entered by permit. Scientists rushed to study these strange new, yet old, birds, and the Ornithological Society of New Zealand changed the name of its bulletin to *Notornis* in honour of the discovery.

But just because the last pocket of takahē had been found didn't mean the birds were out of the woods. At first, 'conservation' of the birds involved just leaving them alone in their protected area, hoping that they would continue to breed and flourish naturally. It quickly became evident takahē numbers were falling, however. Was the takahē once again headed for extinction—for the third and final time?

No one was sure why takahē continued to decline. Some blamed the weather—but surely a bird used to living in an alpine environment could survive a few bad winters? Another likely culprit was red deer, introduced to Fiordland in the 1920s. These large animals didn't kill or disturb the takahē themselves, but they did compete with them for food, eating the sweet tussocks that the birds needed for nutrition. And there were suspicions that stoats were killing takahē chicks and even adults.

A secret plan was developed to try to breed takahē in captivity. A Wairarapa farmer and bird-breeder, Elwyn Welch, was recruited to train bantam hens to incubate and raise pūkeko eggs and chicks. Welch and his special bantams travelled down to Fiordland to collect four takahē chicks from the Murchison Mountains and take them to his farm at Mount Bruce. One of the chicks died, but the others grew into adult birds. To celebrate the success, the takahē were put on public display in 1960, and more than 13,000 people came to view these magical birds which so few members of the public had ever seen.

More adult takahē were brought to Mount Bruce in the 1960s and 70s, and slowly the captive-breeding programme started to produce results. But just having the birds at one other location wasn't enough, especially when the Murchison Mountains population continued to decline rapidly.

A captive-breeding centre was established at

Burwood, near Te Anau, in 1985. Here chicks were hatched and raised using model adult takahē and hand-puppets so they did not become attached to or reliant on human handlers. In the late 1980s and early 90s, birds were translocated to predator-free islands around the country: Maud Island in the Marlborough Sounds, Mana Island and Kapiti Island north of Wellington, and Tiritiri Matangi in the Hauraki Gulf / Tīkapa Moana. Today they are still thriving on Tiritiri, and there are also populations on Motutapu and Rotoroa islands—all within an easy ferry-ride from Auckland city, and a far cry from the remote mountains of Fiordland. Birds have also been translocated to 'mainland island sanctuaries' in the North and South islands, and a second wild population was established in the Kahurangi National Park in northwest Nelson in 2018.

In October 2019, Minister of Conservation Eugenie Sage announced that the takahē population now numbered more than 400, after a record breeding season. Birds are now spread across two 'wild' sites and eighteen sanctuaries in the North and South islands. Despite dire predictions in the 1970s that the original wild population would eventually die out, takahē have also survived in their stronghold in the Murchison Mountains. For now, there is hope.

Ka tū te moho, Kia ora ake anō.
The takahē stands, in order to live again.

Tecomanthe speciosa

12. The rarest plants in the world: *Tecomanthe speciosa, Pennantia baylisiana*

Thomas Cheeseman was excited. Here before him lay the opportunity to explore a wild island few Europeans had ever visited, and it was all his. It was the kind of opportunity a botanist dreamt of.

Great Island didn't look very promising from the sea, he had to admit: sheer rock walls rising out of the ocean, with only a rough coating of greenery visible high up. There looked to be only a few scrubby trees and flaxes . . . but Cheeseman thought there just might be something special tucked away. He had his notebook and pencil handy, and he was determined to do the best job he could in the few hours before the government steamer *Stella* had to be on its way again. It was on the last part of its journey back to New Zealand from the Kermadec Islands.

The party landed on a rocky boulder beach at the head of a deep bay, and Cheeseman started to make his way up the rough cliff at the head of it. The climbing was steep and rugged, but Cheeseman was fit and fuelled by the adrenaline of discovery. Despite the island's appearance from the sea, he was soon finding plenty of plants to note. Lower down there were ice plants and wild celery, then higher up tea tree and bunches of flax. Then—excitement!—a plant he had never seen before, a new type of *Pittosporum*. That one, he thought, he would name after the captain of the *Stella* who had so kindly let him go ashore. There was a new type of fern, growing abundantly on the clifftops. And then a new *Coprosma*. It really was almost too much to take in!

The two hours on the island passed in the blink of an eye. Cheeseman returned to the ship with his notebook and pockets full of specimens. He had recorded 82 species of flowering plants and ferns, and he was sure that this was just the tip of the iceberg. If only he could come back one day, with more time, who knew what else he might find?

New Zealand is a country of wild islands. Dotted around the coasts of the three main motu are many more, large and small. Some are outposts of the mainland, cut off by the rising sea in prehistoric times, and others are volcanoes that have erupted out of the sea. With the encroachment of human settlement and introduced pests, many of these islands have become refuges for rare and endangered species, the sea acting as a protective moat to keep out both animal and human invaders.

Among these special islands is the group known as Manawatāwhi / Three Kings Islands. This group of thirteen islands lies around 55 kilometres northwest of Cape Reinga / Te Rerenga Wairua, and is a scattering of rocky outposts running roughly east to west. The largest is Manawatāwhi / Great Island, while the other main islands are Ōhau / West Island, Oromaki / North East Island and Moekawa / South West Island. The islands are significant to northern Māori as a stepping stone on the final journey of the spirits of the dead to the underworld. Manawatāwhi means 'last breath', and it was here that spirits who had jumped into the sea at Cape Reinga / Te Rerenga Wairua surfaced to look back to their homeland a final time, before plunging underwater and voyaging back to the homeland of Hawaiki.

Despite their proximity to the mainland, the islands have steep sides and are constantly lashed by the ocean swells. Because of this, they have seldom been visited by Europeans, some of them only a handful of times. This inaccessibility and apparent hostility has enabled two of the world's rarest plants to survive there.

Manawatāwhi / Three Kings Islands is one of only two places in Aotearoa which retains the name given to it by Dutch explorer Abel Tasman. (The other is Cape Maria van Diemen, to the southwest of Cape Reinga / Te Rerenga Wairua.) The islands were Tasman's last sight of Aotearoa after he had sailed up the western coasts of the two main islands. The Dutch were desperate for fresh water as they hadn't found a safe place to stop further south, so Tasman anchored off Manawatāwhi / Great Island and tried to send boats ashore. The Dutch visit occurred on the day before the Christian festival of Epiphany, which celebrates the arrival of the three kings bringing gifts to the baby Jesus so, as Tasman recorded in his journal, 'to this island we gave the name of Drie Coningen [Three Kings] Island'.

Tasman's men saw a fall of 'good fresh water, coming in great plenty from a steep mountain', but could not find a place to land safely. On the next day, a second boat was sent ashore but also could not land, 'seeing that the sea was everywhere near the shore full of hard rocks without any sandy ground, so that they would have greatly imperilled the men and run the risk of having the water-casks injured or stove in'. The Dutch saw an area of land being farmed near the waterfall—'square beds looking green and pleasant, but owing to the great distance they could not discern what kind of vegetables they were'—and several groups of Māori, who appeared as giants to the malnourished sailors.

> They saw in several places on the highest hills from 30 to 35 persons, men of tall stature, so far as they could see from a distance, armed with sticks or clubs, who called out to them in a very loud, rough voice, certain words which our men could not understand; that these persons, in walking on, took enormous steps or strides.

The Dutch gave up trying to land and headed north in search of more-hospitable islands.

Groups of Māori continued to live on the islands on and off until around 1840, after which the Kings remained uninhabited and largely unvisited. One of the only times they hit the headlines—literally—was in November 1902. The steamer *Elingamite*, carrying 136 passengers and 58 crew from Sydney to Auckland, along with 52 boxes of silver and gold coins, ran into Ōhau / West Island in a thick fog. Passengers and crew scrambled for the lifeboats, but there was much confusion. Some of the boats made it ashore to Ōhau / West Island and Manawatāwhi / Great Island, but others rowed off into the fog, never to be seen again. The first officer, Len Burkitt, navigated his boat down the east coast of the North Island to Houhora, north of Kaitaia, to raise the alarm, and eventually 41 crew and 108 passengers were rescued.

The captain, Ernest Atwood, was charged with navigational negligence and given a fine, and he had his master's licence suspended. However, nine years later a new survey of Manawatāwhi / Three Kings Islands showed that the islands were in fact three-quarters of a mile (1.4 kilometres) west and 2.75 miles (5 kilometres) south of where they were shown on Captain Atwood's charts. The inquiry was reopened, and Atwood was acquitted of all the charges against him.

The islands were purchased by the government from Ngāti Kuri in 1908, then declared a sanctuary in 1930. In 1956, that status was changed to 'nature reserve' in recognition of the unusual animals and plants that live there. Today, the islands are home to large colonies of seabirds, species of rare skinks and land snails, and other invertebrates, including the giant centipede *Cormocephalus rubriceps*, stick insects, wētā and spiders.

The first botanist to visit the islands and record the rare and unusual species which lived there was Thomas Cheeseman, the curator of the Auckland

Institute and Museum. Despite spending only a few hours ashore, on the way back to New Zealand from the Kermadec Islands in 1877, Cheeseman wrote a paper describing the plants and birds he saw there, providing the first written record of the environment on Manawatāwhi / Great Island:

> Seen from the sea, the aspect of the island is bleak and forbidding in the extreme. Black and rugged cliffs, bare of vegetation and often several hundred feet in height, form the greater part of the shore, and up against them a heavy surf continuously rolls, the spray dashing far up their sides. The summit of the island . . . gave little promise of anything but a very scanty vegetation.

Once ashore, however, Cheeseman found a large number of plants, several of them previously unrecorded species. His list of 82 included five species new to New Zealand and three 'new to science'—not a bad haul for such a short visit. He also noted hawks, tūī, quail (which he misidentified as being the extinct koreke—see chapter 4), an abundance of red-crowned kākāriki and an unusual-looking island form of bellbird, with a paler belly and a different song to the mainland species. Two years later Cheeseman was back for another quick visit, aboard the government steamer *Hinemoa*, along with his new wife Rosetta, who helped him to collect specimens. This time he saw and listed another 60 species.

Cheeseman finished his account of his 1877 visit to islands by saying that, due to their inaccessibility, 'Probably many years will elapse before the peculiar plants and birds are in any way interfered with by human residents'. Human—no. But there were other residents of the islands who were quietly damaging the pristine environment. By the early 1900s there were hundreds of goats living on Manawatāwhi / Great Island, eating whatever they could reach. Māori living on the islands following European contact may have kept goats and pigs there, but it seems most likely that the population sprang from four goats put ashore in the 1890s as a potential food resource for castaways. When they were finally culled in 1946, hunters shot 393 of them. Given the goats' greedy appetites, it's little wonder that when a party of botanical scientists came ashore in 1945 the main vegetation remaining on Manawatāwhi / Great Island was hardy, wind-shaped kānuka trees and scrub.

But as the group explored they found several surprises: four more plant species which had never before been recorded. Of two of these species, it appeared that only a single plant of each had survived the goats—making them the rarest plants in the world.

Geoff Baylis had been studying botany at Auckland University College in 1934 when he got his first opportunity to visit Manawatāwhi / Three

Kings Islands, spending two half-days ashore on an Auckland Museum expedition. After completing his PhD in England and serving in the navy in World War II, Geoff returned to New Zealand and started lecturing at Otago University. In late 1945 he got another chance to go to Manawatāwhi / Three Kings Islands with museum scientists, and this time he spent a week exploring the flora of Manawatāwhi / Great Island. It was long enough for him to find these two fascinating new plant species, both on the brink of extinction.

One was a sturdy multi-trunked tree with very large, broad, glossy leaves with curled edges. Just one tree was found, about 3.5 metres high, branching into four trunks. It was precariously positioned on a steep scree slope, 'facing the sea and about 700 foot [215 metres] above it'.

In 1997, Baylis looked back on this 1945 expedition and remembered finding the tree:

> The last little grove that I investigated lay near the highest point of the island down a scree of boulders about 200m above the sea. I was drawn to it by what looked like a karaka. I was soon gazing upon it in disbelief . . . But this was no karaka—its leaves were larger and recurved strongly in the sun, its bunches of small green flowers sprang from the bare branches below the leaves and there were no big berries—indeed none at all.

Walter Oliver, a retired museum director who had also been on the 1934 expedition, named the tree *Plectomirtha baylisiana* after its discoverer. It is now classified as *Pennantia baylisiana*, as it is related to mainland kaikōmako species, and is also now known as Three Kings kaikōmako or kaikōmako Manawatāwhi. Baylis thought that the tree was 'not in the best of health because of insect damage', and that 'propagating this lone and sterile tree . . . seemed urgent'.

The other interesting discovery made by Baylis in 1945 was a strange vine, with a thick, twisty stem and large glossy leaves, growing up a large kānuka tree near the Tasman Stream on Manawatāwhi / Great Island. Once again Baylis could find only one plant, although 'all suitable habitats were accessible and were examined'. Oliver named it *Tecomanthe speciosa*, the species name meaning 'handsome'. It was the first time a vine from the genus *Tecomanthe* had been found in New Zealand territory; other members of the group are found in Australia, New Guinea and elsewhere in the Pacific.

Both these plants seemed to be the very last ones of their type. Once the goats were gone, it was hoped that they would continue to survive in their natural habitat; but to ensure the continuation of the species, it was decided to try to grow new specimens in New Zealand.

When Baylis came back to the island in 1951, the *Tecomanthe* vine was 'thriving', so he took cuttings from it and brought them back to the Department of Scientific and Industrial Research (DSIR) base in Auckland, where a few years later seeds were successfully produced and germinated. The attractive vine, which produces bunches of drooping, trumpet-shaped, cream-coloured flowers in autumn and early winter, as well as large green seed pods, has now become a popular garden plant, its survival ensured by its presence in many frost-free gardens around New Zealand.

Pennantia was harder to propagate: Baylis also brought back a shoot from its base, which grew well once he planted it in a 'damp, sheltered place' in his Dunedin garden. While waiting for his shoot and other cuttings to grow at the DSIR and the Duncan and Davies nursery in New Plymouth, Baylis wrote:

> I asked George Smith the chief propagator at New Plymouth what I might do to provide better cuttings. 'Cut the tree down' he said, and while I shuddered at the thought he explained that he was confident about rooting shoots from the stump. But would there be any? Well, the tree had four trunks so I dared to sever one. A year later the shoots were there, the Naval launch on which I was a guest gave them a quick passage to New Plymouth which happened to be its next port and Mr Smith soon placed the survival of '*Plectomirtha*' [*Pennantia*] beyond doubt.

Growing the tree from seed proved harder, as the sole remaining specimen in the wild produced only female flowers, and with no male tree to pollinate it, it could not produce viable fruit. However, plant scientist Dr Ross Beever from Landcare Research Manaaki Whenua managed to find a way to encourage it to produce seed which could develop and grow properly, and now hundreds of new plants have been grown this way.

In August 2019, more than 200 *Pennantia* seedlings were donated by Manaaki Whenua and Canterbury Museum to Ngāti Kuri, the iwi whose rohe (tribal area) includes Manawatāwhi / Three Kings Islands. The seedlings were planted around the Waiora Marae at Ngātaki, north of Houhora on the way up to Cape Reinga / Te Rerenga Wairua. For Ngāti Kuri, it was the return of long-lost taonga to their land.

The original *Tecomanthe* and *Pennantia* plants continue to survive on Manawatāwhi / Great Island, though the vine has been a victim of the success of the island's forest regeneration. The bush around it became too thick for it to thrive, and it failed to flower for 50 years. However, careful pruning of plants around it and training of the vine has resulted in it bouncing back, and six more plants have been established from cuttings from the original. It has flowered several times but not yet produced any fruit.

What was once the world's sole *Pennantia* tree still

stands alone on its windblown cliff—but hopefully not for long. In March 2020 some of the mainland-grown plants will be returned to the island and planted near it. While this individual tree remains vulnerable to a natural event such as a storm, slip, fire or drought, which could finish it off for good, at least the long-term survival of the species seems assured.

Mokohinau stag beetle (*Geodorcus ithaginis*),
left, and Canterbury knobbled weevil
(*Hadramphus tuberculatus*)

13. All the small things:

Vanished lizards, frogs, worms and insects

It was a fine day, and Fremming Sandager had a few precious hours to himself. His life was determined by his role as assistant keeper of the Mokohinau Islands lighthouse, and much of every day was spent tending the light and keeping it aglow. The wick of the oil lamp needed to be trimmed regularly, the glass lenses polished and the revolving mechanisms wound every hour or two to keep the light turning. Then there were farming duties, working with the keepers' little herd of livestock, and of course the record-keeping.

Not that Fremming needed much time off—it wasn't like there was anything to do on the tiny island, cut off from civilisation by so many miles of sea. Before this he had been at Tiritiri Matangi, just a short ship-ride from Auckland. Here, the supply ship came only once every three months—if they were lucky.

His favourite way to pass the time when not on duty was pursuing natural-history specimens. The islands might be bereft of organised entertainment or other trappings of civilisation, but they were rich in birds, bugs and lizards. The seas around the rocky shores teemed with interesting fish. Fremming was slowly building a collection of fish skins and specimens of the living things that crept and crawled about on the island.

One of his favourites was a large, shiny black beetle, which he had found on the island they called Lizard Isle because there were so many shiny brown-spotted skinks there. The beetle had impressive 'antlers' like a European stag beetle, and Fremming was sure it would have put up a fight if he had found it alive. Unfortunately it wasn't complete, and try as he might he hadn't yet been able to find a live one.

Maybe today would be the day. Wilson was on duty, and he had said that Fremming could take the rowboat over to the island for another look. Fremming would so dearly love to carefully pin another beetle into his case and preserve it forever.

When talking about the lost wonders of Aotearoa, the first things that spring to mind are the wonderful birds that have become extinct—especially iconic species such as the moa and huia. When thinking about our conservation success stories and those treasures that have been snatched back from the brink of oblivion, the charismatic kākāpō or the cute-as-a-button black robin come to mind. But what about the small things, the creepy-crawlies, the lizards, the insects, the invertebrates that shuffle through the leaf litter, crawl up tree trunks and busy themselves in their work of breaking down plant matter into soil? These too are wonders of Aotearoa's biodiversity, and a large number of them are just as threatened as the headline species of highly visible, photogenic and 'marketable' rare birds.

Today, Aotearoa has a rich and diverse fauna and is home to thousands of species of insects, invertebrates, lizards and frogs, many of them unique to these islands. But we may never truly know what we have lost since the arrival of humans.

Today New Zealand is home to three types of *Leiopelma* frog, all of which are threatened or at risk: Archey's frog (*L. archeyi*), Hamilton's frog (*L. hamiltoni*) and Hochstetter's frog (*L. hochstetteri*). This ancient type of frog is unusual in that most species lay their eggs on wet ground, not in water, and they do not have a tadpole stage. Subfossil bones found around the country suggest that there used to be at least three more kinds of native frog; the Aurora frog (*L. auroraensis*), remains of which were found in a cave in Fiordland; Markham's frog (*L. markhami*), identified from bones found in the Kahurangi National Park in northwest Nelson, and estimated to be up to 6 centimetres long; and the Waitomo frog (*L. waitomoensis*), which was almost twice that size. No one knows when or why they became extinct.

Beneath our feet, New Zealand is also home to more than 170 different species of native earthworm. In our gardens we mostly see the introduced lumbricid worms, brought here to improve soil conditions in the new farmlands. The majority of native worms are adapted to live in forest-soil environments. Several of these species are very large, growing up to more than a metre long, and they were harvested by Māori to use as bait for fishing, and also as a food source. In 1902, ethnologist Elsdon Best wrote about the Māori tradition of eating worms:

> To cook these worms some water is placed in a bowl and rendered warm (not hot) by means of hot stones. The worms are then cast into the water and allowed to remain there for some hours. Before long (before the sun sets) the worms will have become dissolved . . . Some cooked puwha (greens) is added to the mess and a prized dish is ready; the gods who live for

ever would smile at the sight of it . . .

The two most prized kinds [called kurikuri and whiti] were reserved as food for the chiefs. The sweet flavour (tawara) of those kinds is said to remain in the mouth for two days. I cannot speak from experience.

Around 30 of these native worm species are considered at risk, although the majority are classified 'data deficient'—scientists just don't know enough about them to decide whether they are endangered or not. One species is thought to have become extinct: *Tokea orthostichon*, apparently first collected by Austrian naturalist Ludwig Karl Schmarda at Maungarei / Mount Wellington in Auckland in 1861. It is known only from a handful of museum specimens and has not been seen for more than 150 years, so it is now believed to have vanished from the world.

Another mysterious former resident of New Zealand is also known from one museum specimen. A single stuffed giant gecko came to light in a museum in France in the 1970s, starting a flurry of discussion as to whether the giant kawekaweau lizard of Māori legend might have been based on fact.

Māori called native lizards—both the smooth-skinned skinks and the velvety geckos—mokomoko. They were associated with the atua Whiro, who personifies evil, darkness and death, so they were feared and considered a bad omen. There were also legends of giant lizards or ngārara, which were a type of taniwha or monster. Among them were kawekaweau, a large gecko. East Coast ancestor Kahungunu was said to keep one in a special bowl, using it to scare approaching enemies.

Stories of the kawekaweau lingered on once European settlers arrived here, and were accepted as fact. In 1870, naturalist and collector Walter Buller wrote:

> The Kawekaweau, a beautiful striped lizard, sometimes attaining a length of two feet [61 centimetres] . . . was formerly abundant in the forests north of Auckland, and is still occasionally met with. Mr. F. E. [Frederick] Maning, of Hokianga, recently obtained possession of a pair of live ones, but unfortunately for science, one of them was devoured by a cat and the other made its escape.

There were also other reports: in 1872 Captain Gilbert Mair told the Auckland Institute of

> the existence of a large forest lizard, called by the Maoris [*sic*] kaweau. In 1870 an Urewera chief killed one under the loose bark of a dead rata, in the Waimana valley; he described it to me as being

> about two feet long, and as thick as a man's wrist; colour brown, striped longitudinally with dull red.

In 1894, however, Buller wrote that the lizard had become rare or extinct 'within the last thirty years, when the remnant of its race succumbed to wild pigs and other natural enemies'. This timing tallies with a report from 1904, in which Archdeacon Philip Walsh reported stories of 'a large lizard' which once lived on the Waoku Plateau south of the Hokianga Harbour in Northland:

> So far as I am aware, no specimens have been captured; or, if they have, they have not been preserved. A dead specimen was, however, washed down the Waima Creek . . . about thirty-five years ago [in the 1870s] . . . when it was seen by several European visitors, and was recognised by the Maoris [*sic*], who were much frightened at its appearance. Being in a partly decomposed condition, however, no attempt, I believe, was made at preservation.

He also shared a report by a local farmer of seeing a large lizard 'about 18 in. [45 centimetres] long of a yellowish colour' on the banks of a creek in the area in the 1850s. There seem to have been no later sightings.

However, in the mid-1980s, a pair of North American gecko researchers, Aaron Bauer and Tony

Russell, started looking into the provenance of a specimen found in the Natural History Museum of Marseille in the south of France. From examining its scales and (incomplete) skeleton, they concluded that it was a *Hoplodactylus* gecko, one of a group of nocturnal brown-skinned geckos found only in New Zealand. The sole specimen was around 37 centimetres long, with reddish-brown stripes—similar to Mair's description—which would make it 50 per cent bigger than any other known *Hoplodactylus* species. It was given the name *Hoplodactylus delcourti* or Delcourt's gecko, after the curator of reptiles and amphibians at the museum, Alain Delcourt, who had found it in the museum basement in 1979.

Very little else was known about the huge lizard—neither when or where it was collected, or by who. It was believed to have been at the museum since the 1860s, and to have come from New Zealand. The specimen, while mounted in a standing position, was found to have been constructed around a skull, all four legs and one other bone, but to be missing its spine and tail bones. The lizard had been on public display in the past without anyone really knowing what it was.

The Natural History Museum of Marseille was founded in 1819, and the researchers who examined the stuffed gecko after its rediscovery think that it was most likely brought there between 1833 and 1869. They guessed that it could have been collected by early French visitors to the Far North.

We will never know who found Delcourt's fabulous specimen and brought it back to France. The mystery of whether or not the kawekaweau was still alive little more than 100 years ago will also remain unsolved, but it seems certain that such a large lizard could not have gone unnoticed for another century. While it may once have walked the forests of Aotearoa, kawekaweau seems to now exist only in legend.

While ancient insect remains have been found fossilised in rock or preserved in hardened tree sap or amber, more-recent deaths leave no trace: insects have no bones that could be dug up. Insects can also be hard to see and catch, and early observations of them here were brief.

Māori were aware of and named a large number of different insect species, some of which were used as food, such as the nutritious grubs of the huhu beetle (*Prionoplus reticularis*), or tunga rere. From the first European settlement, natural historians such as William Colenso collected specimens and published articles on them. Auckland Museum holds what it believes to be the oldest insect specimen in New Zealand: a giant wētā caught by Colenso in 1838.

With the destruction of native forests and the spread of human habitation, it seems likely that many species of insect quietly became extinct before anyone even knew they existed. We know that there was a particular type of louse (*Rallicola extinctus*) that lived only in the feathers of the huia; when the bird became extinct, so did the louse. (Its existence was discovered long after it had disappeared from the world, through specimens found among the feathers of stuffed huia and their skins.) *Mecodema punctellum* was a large black flightless beetle, about 4 centimetres long, which was found only on Stephens Island (see chapter 6). It was last seen in 1931, and subsequent surveys of the island have failed to turn up any other specimens.

Today, many of New Zealand's unique species of insect are under threat, and certainly in need of further study. But there are some that we know have definitely disappeared—and one that fortunately came back from the dead.

In December 2004, University of Canterbury Master's student Laura Young was studying the native *Aciphylla* speargrass at Burkes Pass, on the edge of the Mackenzie Country near Lake Tekapo in the South Island. Young and her field assistant Sarah Luxton came across a large flightless weevil crawling about the base of one of the plants. They knew of the existence of the Canterbury knobbled weevil (*Hadramphus tuberculatus*), and also knew that it had last been seen in 1922 and was believed to be extinct. However, once an entomologist (a person who studies insects) got a look at a specimen it was official: *Hadramphus tuberculatus* was not extinct at all. The trouble was that it seemed to be restricted to this one 10.5 hectare patch of speargrass—and so was at high risk of being accidentally wiped out.

When it was first described in 1877, the knobbled weevil was said to 'abound' on the plant known as wild Spaniard—an old name for the spiky *Aciphylla* plants once found widely on the Canterbury Plains and foothills before their conversion to pastureland. Canterbury knobbled weevils grow to around 1.5 centimetres long, and as their name suggests they have grey-brown bodies covered in bumps and lumps called tubercules. Four types of *Hadramphus* weevils are found in New Zealand, the others all now restricted to offshore islands: the Poor Knights in the north, the Chathams, and islands around Fiordland and Rakiura / Stewart Island. The southern variety, *Hadramphus stilbocarpae*, was once found on Taukihepa / Big South Cape Island, but the population there was destroyed by the same rat infestation that drove the South Island snipe and Stead's bush wren to extinction (see chapter 8). All four species, which are largely nocturnal, are considered threatened and are protected by law. Their natural predators are owls like the ruru and, in the past, the whēkau (see chapter 4), but today their main threats are rats, mice and hedgehogs, which eat the adult weevils, and destruction of their habitat and the plants they eat.

Adult *Hadramphus* weevils feed on the leaves, flowers, pollen and seeds of the plants they live on (at their immature larval stage, the weevils live in the soil and eat the plants' roots). Canterbury knobbled weevils have a particularly close relationship with *Aciphylla* speargrass, and have adapted to eat its tough and spiky leaves. This means that as the speargrass has become rarer, the weevils have disappeared alongside it.

After the rediscovery of the Canterbury weevil, further searches found that there were up to 200 living in the reserve at Burkes Pass, and another small group nearby. However, monitoring has shown that the numbers of the weevils are decreasing, and despite other searches, no more weevils have been found elsewhere. The Canterbury knobbled weevil is currently classified as 'nationally critical' because it is found only in this one area.

The ongoing survival of the knobbled weevil, which is considered one of the top-ten most at-risk species by the New Zealand Endangered Species Foundation, relies on keeping the existing insects safe from introduced predators, and encouraging the growth of the *Aciphylla* speargrass. Wallabies and hares also eat *Aciphylla* and pigs grub up the roots. The plants are also at risk of being damaged by fire, as happened at the reserve in 2005 when a spark from a bus passing on the nearby road started a blaze. Researchers have been experimenting with captive breeding of the weevils—with some success—and are hopeful that this can be used to boost weevil numbers if the population continues to fall. Translocation of some weevils to another suitable, better-protected site is also being considered, and weevil enclosures are being trialled to protect the insects from predators. There are only about 100 knobbled weevils remaining, but they are not going to be left to slide into oblivion.

It may already be too late for the other insect on the Endangered Species Foundation's top-ten most threatened list, the Mokohinau stag beetle (*Geodorcus ithaginis*). This large, shiny beetle with its distinctive horns is found on only one island in the Mokohinau group in the northern Hauraki Gulf / Tīkapa Moana—a rocky pile known as 'Stack H'. This stag beetle is even rarer than the knobbled weevil, and may already be extinct. A recent expedition to the islands, recorded in *New Zealand Geographic* magazine in September 2019, could find no live beetles at all, and only a few remains of dead ones.

The Mokohinau Islands lie off the Northland coast, about 100 kilometres northeast of Auckland. In pre-European times, Māori came to the islands to harvest muttonbirds, bringing with them little shipmates—kiore—who found plenty to munch on, from birds' eggs to seeds to skinks to a range of invertebrates. There is no doubt that Mokohinau stag beetles were to its liking.

In 1883, a lighthouse was built on the largest island, Burgess Island (Pokohinu), to signal the northern entrance to the Hauraki Gulf. Its extremely isolated location made it an unpopular posting. The first assistant lighthouse keeper, Andreas Fremming Sandager, passed his time when not on lighthouse duty collecting specimens of the island's flora and fauna, including 130 species of beetle. He was the first to scientifically describe the stag beetle, after finding a female and the remains of a male on Lizard Isle. This previously largely untouched island was invaded by kiore in 1977, and no beetles have been found there since.

After the lighthouse was built, as on many offshore islands around New Zealand much of the remaining forest was burned off or chopped down and farming was attempted. This destruction of natural habitat, combined with the kiore, meant that the stag beetle died out on the main islands of the Mokohinau group and became restricted to one rock stack with an outcropping of wind-blown pōhutukawa and low-growing ice plants and coastal tussock grass. Kiore have now been eliminated from the other islands of the group and the forest is being left to regenerate, but the stag beetle remains isolated on Stack H.

The male Mokohinau stag beetle grows to around 3 centimetres long and has distinctive antler-like jaws, with a horny spike on top, which it uses to fight other males when defending territory or competing for a female. The females are smaller and do not have enlarged jaws. The beetles are flightless, and they can be hard to find as they are largely nocturnal, burrowing in leaf litter and hiding during the day.

They have not been seen many times since Sandager's original description: further specimens were collected in 1902, then none were recorded for more than 80 years. A male and parts of a female were found in 1984 by members of a University of Auckland field club researching the lizards of the Mokohinau Islands. In the early 1990s, Department of Conservation expeditions to the islands discovered a handful of beetles living in the pōhutukawa litter on Stack H. The most ever seen on one expedition was nine.

Because the island the beetle lives on is so small, the species is incredibly vulnerable to any significant change in its environment: a stray rat, a fire, a bad storm. Sadly, one of the biggest threats to the survival of the Mokohinau stag beetle is humans. Just like the bird collectors of the 1800s, who sought endangered birds as specimens for their private museums, insect enthusiasts today pay big money for rare insects, dead or alive. No one knows whether the last beetles—which are protected by law—have been poached, but it is a possibility.

If more beetles could be found it might be possible to breed them in captivity and increase the species' range by releasing some on to other islands. However, so little is known about the beetle—and they are so hard to find—that it is impossible to know if they will survive, or if they will join the list of lost wonders of Aotearoa.

Kākāpō (*Strigops habroptilus*)

14. Parrot of the night: Kākāpō

Richard Henry stood up and dusted his hands on his overalls. It was fine now, but there was bound to be rain later. Here on Resolution Island, tucked into the entrance to Dusky Sound, it rained 200 days of the year—sometimes, it seemed, all of those in a row. No wonder Captain Cook had called a nearby inlet of the sound Wet Jacket Arm.

Richard had been squatting, looking at a strange dip in the ground where the grass was flattened as if it had been trampled down by tiny feet. It looked like the kākāpō were getting ready to breed again.

He could hear them at night, making their strange calls—not their usual cackles and parrot-like *skrark* but the resonant booming sound the males made to attract their mates.

Richard had been closely watching the birds and making notes about them in the decade that he had lived in the bush. For the past few years he'd been acting as caretaker of the island reserve. It was his own special place—he and his assistant Andrew Burt were the only ones who went there, rowing across from the little house they'd built on Pigeon Island. Richard himself had caught most of the hundreds of flightless birds now kept safe on the island, and he knew their ways. The big men up in Wellington might think his observations and theories were unscientific, but he knew the kākāpō better than anyone. One day they would listen to him and find out how much knowledge he had accumulated about these strange birds.

Then something on the ground caught his eye, and he crouched down again. It was a dropping, but not all soft and seedy like the little mounds the kākāpō left behind. This was smaller, darker—a thin tube.

Richard poked at it with a stick. He could see little pieces of feather, tiny pieces of bone.

He knew what this meant. Stoats had come to his precious island, and the birds there were now on borrowed time.

Of all the weird and wonderful birds which once roamed the islands of Aotearoa, the kākāpō has to be one of the strangest. It is the largest parrot in the world, and the only one which is both nocturnal and flightless. Adapted over millions of years to scrabble about the forest floor in the darkness, the kākāpō is completely and utterly unique—and it was nearly lost to the world forever.

Today, the kākāpō is one of New Zealand's greatest and most visible conservation success stories. News items regularly update readers on the progress of each breeding season or when an adult bird dies. During the winter of 2019, an outbreak of a kākāpō respiratory disease caused by a fungus was covered closely by the media—as were the celebrations when the population rose to over 200 birds after a successful breeding season. The species even has its own 'ambassador', an adult male bird called Sirocco, who tours zoos and sanctuaries around the country to raise awareness of his species. The kākāpō is still critically endangered, but today huge amounts of time and money are being spent to make sure it doesn't slide quietly into extinction.

It wasn't always this way. In the nineteenth century, with the spread of human settlement and introduced pests, numbers of kākāpō dwindled throughout the country. Its habitat was destroyed, and rats and stoats preyed on its chicks and eggs. By the 1930s, many people thought that the species was extinct. No one even thought to look for them for 30 years. The kākāpō was all but forgotten.

This is the story of how the kākāpō was all but lost, then brought back from the brink through sheer hard work—and a lot of luck.

Today we think of the kākāpō as a bird found only in remote areas of the South Island, but before the arrival of Europeans in Aotearoa it was found throughout both the North and South islands. Māori called the bird kākāpō because it was a parrot like the day-flying kākā and kākāriki, but of the night, te pō. It was also known as tarepō. When the first Polynesian voyagers arrived in Aotearoa, the kākāpō was widespread. Māori knew the birds and their habits well—hundreds of years before European scientists figured it out, they recognised that the kākāpō do not breed every year, and that the male birds use a special dance to attract mates.

While the birds were well camouflaged and nocturnal, they were still easy to catch and kill during the breeding season. The loud calls of the male kākāpō drew Māori hunters to them in the dark, and kurī could sniff them out. If a number of male birds were out 'booming' to attract mates, the hunters would take the young 'sentry' bird first, so that the others would not know they were being attacked. Many birds were collected and preserved in their own fat in tōtara bark containers called pātua, or in bags made from the blades of kelp, for consumption over the year. The birds were snared and trapped in pits, too, or taken from their nests by dogs. Māori also harvested and ate the birds' large eggs, which are about 5 centimetres long.

Kākāpō skins and their striking bright green and yellowish feathers were highly prized for making cloaks and other clothing. The pepeha 'Te kākahu ō mea he kākāpō', indicating a person of rank, means 'His clothing is indeed that of the kākāpō'.

Today, thanks to the efforts of conservationists since the 1970s, most New Zealanders know what a kākāpō looks like. They might not have seen one 'in the feather', but their images are widely broadcast in the media. With their quizzical expressions, stumpy wings and bright-green rugby-ball-shaped bodies, kākāpō are highly photogenic and have become one of our most iconic birds. What many people might not appreciate, however, is how big they are—up to 64 centimetres long and weighing up to 4 kilograms—larger than a domestic cat. The other thing that photographs and videos cannot show you about the kākāpō is their unusual sweet and musty scent, which has been described as smelling like the inside of a violin case. All in all, they are a truly unique package.

Because of their nocturnal, bush-dwelling habits, kākāpō were not observed by curious Western eyes for some time. Neither James Cook's naturalists in the 1700s nor the sailing-ship crews that followed in their wake in the early 1800s mentioned seeing one. It was first described by a European observer in 1845, when ornithologist George Gray of the British Museum in London was sent the skin of a bird captured in Dusky Sound. He gave the bird the Latin name *Strigops habroptilus*, meaning owl-faced bird with soft feathers.

David Lyall, a surgeon on the British survey ship *Acheron*, was the first to write a scientific paper on the unusual bird. The *Acheron*, a state-of-the-art paddlewheel steamship, was sent to New Zealand in 1848 to record the new colony's coastline and generate new and more accurate charts of Cook Strait and the South Island. The crew of the *Acheron* encountered kākāpō in Doubtful Sound, and Lyall took several birds on board to study. Captain John Lort Stokes recorded in his journal:

> . . . for want of a proper cage, a very handsome Kakapo and in excellent health and condition, was imprisoned in the Dispensary—So long as daylight continued the animal remained in a corner but on the approach of night became quite frisky—as all night birds invariably do. Having clambered upon the shelves among

> the jars and bottles he . . . 'skoffed' some dozen pills and with the inner man thus satisfied, flew through the skuttle and was drowned.

Not a bad effort for a flightless bird!

Lyall did manage to keep one of his specimens alive to 'within six hundred miles [965 kilometres] of England (when it was accidentally killed)'. Captain Stokes gave another bird to Major George Murray of the 65th regiment at Wellington, who kept it as a pet in his garden, 'where it was fond of the society of the children, following them like a dog wherever they went'.

In a paper presented to the Zoological Society in London in 1852, Lyall said that while the bird could still be found in inland, mountainous areas of the North Island, it lived 'in considerable numbers' in the damp sounds of Fiordland. Lyall noted that the birds were seen to 'fly' only once, but his description of the flight suggests that it was more of an ungainly flop between the branches of a tree. He also saw them climbing using their strong beaks and claws for grip and balance. Even at this early stage of European settlement, Lyall noted that the kākāpō population was being dramatically reduced by predation by 'wild dogs', adding, 'it is to be feared that if they once succeed in gaining the stronghold of the kākāpō [in South Westland] the bird may soon become extinct'. However, Lyall finished his account by adding, 'Their flesh is white, and is generally assumed good eating.'

More stories of this strange nocturnal parrot came to light as European settlers advanced into the forested interior of both islands in search of resources and land. By the late 1800s kākāpō seemed to be restricted to the remote, forested southwestern corner of the South Island, but in these relatively undisturbed forests they could still be found in abundance. Charles Douglas, an explorer of South Westland, wrote in the late 1890s that the bird had been 'common at one time all over Westland, but is now confined to the country south of Bruce Bay'. Numbers had fallen dramatically since the spread of ferrets and stoats to this part of the country. In 'good kakapo country', Douglas wrote,

> The bird used to be in dozens round the camp, screeching and yelling like a lot of demons, and at times it was impossible to sleep for the noise. The dog had to be tied up or matters would have been worse. It would have been killing and fetching all night long . . .

Dogs, Douglas noted, were fond of eating the birds: 'Few dogs will eat a weka, a kea, or a kiwi, but I never saw one that refused a kākāpō.' They provided sustenance for human travellers and bushmen, too. The young ones were apparently good roasted, while the older birds had to be boiled or cooked hangi-style in the ground because they were too tough. Douglas wrote, 'When very fat, the bird gives almost as much oil as a weka [and] is useful for frying or making cakes or shortbread.'

Some Europeans also kept them as pets, although they didn't seem to survive long in captivity. (Douglas wrote of a group of five kākāpō that he tried to cage together overnight, which fought each other to the death.) Birds were also shot for museums and private collections, but the greatest damage to their numbers was caused by the felling and burning of the forests in which they lived, and the introduction of pests such as ferrets, stoats and weasels in the 1880s. Like many of New Zealand's unique species, the kākāpō entered a decline which sped up as the nineteenth century progressed.

One of the kākāpō's first champions—who saw them as something to be studied and preserved, rather than eaten—was an amateur naturalist, loner and bushman called Richard Henry. Henry was born in Ireland but brought up in Australia, and he came to New Zealand in 1871. After travelling around, he set up a base at the southern end of Lake Te Anau. Henry got to know the birds of Fiordland through close observation. He was the first European to learn of the kākāpō's strange and irregular breeding habits, and to speak out for its conservation.

When Resolution Island in Dusky Sound, Fiordland, was set aside as a reserve for the 'protection of native flora and fauna' in 1891, Richard Henry's name was put forward to be its curator. In July 1894, he and his assistant Andrew Burt travelled to the remote sound and were left to get on with the job. Over the following six years, it is estimated that Henry and Burt captured around 750 birds—not only kākāpō, but also kiwi—and transported them to the sanctuary island. They also sent around 100 other birds to other reserves and private keepers.

However, Henry was ever mindful of the island's proximity to the mainland, and in 1900, his worst fears were realised. Stoats invaded, and keeping the birds safe became an uphill battle. A few kākāpō were moved to Te Hauturu-o-Toi / Little Barrier Island and Kapiti Island off the North Island; the Barrier birds were never seen again, although those on Kapiti lived on into the 1930s. Discouraged, Henry left Resolution Island in 1908, and the fate of the kākāpō was all but forgotten for 50 years.

From the middle of the twentieth century, the government started to pay slightly more attention to New Zealand's native wildlife. A branch of the Department of Internal Affairs was made responsible for the conservation of rare native species and carrying out field work and investigation. The kākāpō was one of the first species to benefit from this increased attention, and in 1958 the first expedition to find out whether the bird even still existed arrived in Fiordland. No one really even knew where to start looking, but after consulting historical records and recent sightings, a group set off up the Tūtoko Valley, near Milford Sound / Piopiotahi. After very little searching, their dog found a large, one-eyed kākāpō asleep under a rock. The bird was photographed, measured and set free. Like the takahē (see chapter 11), the kākāpō has been proved to be still alive.

Further expeditions explored not only Fiordland but also northwest Nelson and the Tararua Range, between Wellington and Palmerston North (Te Papaioea). While more birds were found in the south, kākāpō didn't seem to be living anywhere else.

Initially, several Fiordland kākāpō were captured and taken to the farm of amateur bird-breeder Elwyn Welch at Mount Bruce in the Wairarapa, but they failed to thrive in captivity and soon died. Scientists were frustrated; the birds became stressed when kept near each other, refused food and showed no signs of breeding (although after their death it was discovered that the first five birds taken to Mount Bruce were all males). By the early 1970s, the kākāpō seemed worse off than ever, with no known birds surviving.

In 1973, the man who would be hailed for saving the kākāpō joined the Wildlife Service's field team: Don Merton (see chapters 8 and 9). Under his leadership, in the mid-1970s several more birds were found and caught in Fiordland, including one that was named after the original Richard Henry. They were moved to Maud Island in the Marlborough Sounds, and then to Te Hauturu-o-Toi / Little Barrier Island in the Hauraki Gulf / Tīkapa Moana when stoats invaded Maud just as they had Resolution. But again, all the birds that had been captured were males. The situation seemed desperate. The kākāpō had been found once more—but could it really be saved?

One of the biggest problems facing the team trying to stop the kākāpō from becoming extinct was its unusual breeding habits. It took researchers until the 1970s to figure out that the bird has a 'lek' mating system. This means that groups of males gather together in an 'arena' and call and display to attract mates. The male birds form systems of tracks through the undergrowth to 'bowls', where they make their distinctive booming calls during the mating season. When the Wildlife Service officers figured this out, they knew they had no chance of breeding the birds in captivity.

There was also one other complicating factor. As Richard Henry and other early observers had noted, kākāpō only nest in certain years—and then all the birds breed at once. Its successful breeding is reliant on the heavy fruiting, or 'mast' years of certain trees—particularly beech and rimu. Every few years, following a particularly warm summer, these podocarps produce a bumper crop of fleshy-footed seeds—a signal to the kākāpō that there will be plenty of food to feed their chicks. It is yet another reason why encouraging kākāpō to breed is such a complex process—and why captive breeding wasn't going to work. The Wildlife Service had to do what they could to help the kākāpō to help itself.

Just when things were looking particularly bleak for the kākāpō, there was a breakthrough of sorts. In 1977, Wildlife Service officers searched for the birds on Stewart Island / Rakiura. No kākāpō had been seen there since 1949, but the island was largely predator-free and there were high hopes of finding kākāpō there.

Their confidence was rewarded: in the southern part of the island a large population of up to 200 birds was found, including the first females seen for more than 70 years. But there was another problem. Stewart Island / Rakiura might have been free of stoats, but it was home to a large and hungry population of feral cats, which were making a meal of the kākāpō—literally. In the five years following the discovery, more than half the birds being tracked by scientists were killed by cats. The kākāpō were dying as fast as the Wildlife Service officers could find them.

Getting rid of the cats was considered impossible, so the kākāpō had to move home. A number of birds were translocated to Te Hauturu-o-Toi / Little Barrier Island, then a small island off the southwestern tip of Stewart Island was declared free of possums and weka, and suitable for kākāpō. Codfish Island / Whenua Hou became the new stronghold for the species, with all the remaining birds from Stewart Island moved there, or to Maud or Mana islands, in the late 1980s and early 90s. Birds were also released on Te Hauturu-o-Toi / Little Barrier Island in 1982, and on Anchor Island in Dusky Sound in 2005 once stoats had been exterminated there. Kākāpō were now considered extinct in their natural range in both Fiordland and Stewart Island, but there was a chance they might survive in their new homes.

And so began the modern phase of the recovery of the kākāpō. The tale of the species over the past 30 years has been an up and down one, of alternating triumph and disaster. The Department of Conservation, established in 1987, began to closely manage and study the remaining birds in the wild, equipping them with trackers and observing them throughout their life cycle. Scientists can now predict when a breeding year is likely to occur, and they closely monitor each egg and chick, providing supplementary food, medical care and hand-rearing if required. Artificial breeding techniques are used to ensure genetic variation in the population. But even with the weight of science and technology behind them, DOC staff still have to wait, just like the kākāpō, for the right set of environmental conditions before breeding can occur.

From a low of just 50 birds in 1995, the kākāpō population officially reached 213 in September 2019. The summer of 2018–19 was a bumper breeding season for kākāpō, with all except one of the adult females breeding, and one super-mum laying three clutches of eggs! Of 252 eggs laid, 86 hatched successfully and a new record of 71 baby birds made it through their first five months of life. This means that there are more kākāpō in the world today than there have been for at least 70 years.

One of the challenges the Kākāpō Recovery team faces now is finding enough space for all the birds. Options beyond the current three sites where wild kākāpō currently live (Codfish Island / Whenua Hou, Te Hauturu-o-Toi / Little Barrier Island and Anchor Island) have to be considered.

For now, things look good for the kākāpō. But the birds' long-term survival relies on a lot of intervention and management. Climate change may also be a factor, as it may affect the fruiting of beech and rimu. Whatever happens to the species in the future, it seems certain that this time they will not be left to slip away.

Part 3
Almost lost?

Māui dolphin (*Cephalorynchus hectori maui*)

15. Te whānau puha: The great whales and the Māui dolphin

What struck Jerningham Wakefield first was the smell. Coming ashore from the *Tory*, the thick scent of rotting flesh and hot oil rose from the sunny little bay. So this was Te Awa Iti, or 'Tar White' as the whalers called it. There was a clear stream running through the centre of the little town, but where it met the sea the water was stained red with blood and bobbing with offcuts of whale carcass. There were around twenty houses, mostly built from rough clay with roofs made of reeds, although the head whaler, Dicky Barrett, had a more substantial dwelling made of sawn timber.

It was the Sabbath, but the little settlement was busy. A whale had been towed ashore a few days earlier, and there was no time to waste to get it processed and the oil rendered out of it before it went even more rotten. A few of the whalers, both Māori and Pākehā, were dressed in what Jerningham guessed was their 'Sunday best', but those labouring at the try-works boiling down the blubber were unshaven and uncombed, their clothes covered with dirt and oil. The beach was saturated with oil, and the stench of the cut-up pieces of whale carcass, lying about in the sun, was intolerable.

'Do you always work on Sundays?' Jerningham ventured to ask one of the men as he stirred one of the great iron cauldrons. He wiped a greasy hand across his brow and grinned, revealing a mouth largely empty of teeth.

'Oh, Sunday never comes in this bay,' he replied.

The waters around the islands of Aotearoa were once rich with marine mammals. The rocky shores overflowed with seals and sea lions, and the inshore waters were visited by baleen whales, passing by on their annual migration between Antarctic waters and the Pacific. Further out in the deep water, huge toothed sperm whales fished for squid. Smaller toothed whales, from tiny dolphins to bulky black orca and pilot whales, cruised the harbours and the coastlines. The nutrient-rich waters of the southern Pacific Ocean provided whales and seals with plenty of food, and they were undisturbed for thousands of years.

Whales were considered special animals by Māori, as the offspring of Tangaroa, the atua of the sea. Collectively they are known as te whānau puha—the family that breathes out air. Specific names include tohorā or tohoraha, the southern right whale, and parāoa, the sperm whale. Whales are said to have guided some of the first waka to Aotearoa, and Ngāti Porou on the East Cape draw their lines of descent back to Paikea, who was said to have ridden here on the back of one. Whales were a symbol of bounty, providing food for many people, and symbols of whales were often carved onto the maihi (bargeboards) of pātaka (food storehouses), indicating the abundance within.

Before European whalers arrived and taught them the trade, Māori did not actively hunt whales at sea. However, a stranded whale could be a godsend for protein-hungry tribes. As well as the whales' meat and blubber, their bones and teeth were used for carving ornaments and making weapons such as patu and taiaha. Māori may also have netted the smaller dolphins around the coast, or encouraged larger whales to strand by driving them into shallow water.

It wasn't until the first European ships started to explore further into the southern Pacific Ocean that the real hunting began. Southern whaling started off the British penal colony of New South Wales in the early 1790s, when ships that had been chartered to transport prisoners down under started to make the most of their long journey by filling up with oil for the return trip.

The first whaling ship recorded to call into New Zealand was *William and Ann*, a British ship captained by American Eber Bunker, in early 1792—just twenty years after Captain James Cook had put Aotearoa on the world map. (Cook and his men had reported seeing many whales on their voyages around these islands.) Bunker cruised the coast of New Zealand in search of whales on his way back to England from Port Jackson (Sydney), stopping in Doubtless Bay.

Over the next few decades more and more ships called into the northern ports, especially the Bay of Islands, while on whaling voyages hunting the pelagic (deep-ocean living) sperm whale (*Physeter macrocephalus*). Pelagic whaling was hard, dirty and dangerous work. The massive whales were attacked with harpoons from small boats, then cut up and rendered down for their oil at sea. Whaleship crews would sometimes be away from their home ports for years at a time, so pop-up settlements such as Kororāreka in the Bay of Islands provided a welcome rest, with fresh food, alcohol and entertainment for sale.

The whaling industry reached its peak in the 1830s, when a ban on American whaleships visiting Australian ports was lifted. A flood of New England whalers, who were finding that whales were becoming less plentiful further north in the Pacific, started to voyage down to what became known as 'the New Zealand ground', where many whales could be found to the northeast of the North Island.

The sperm whale was targeted because of the large amount of high-quality 'spermaceti' oil found in a special cavity inside its enormous head. Each whale's 'case' could contain up to 2000 litres of oil, which was mostly used as fuel for lamps or made into candles, in the days before petroleum-based mineral oil. Sperm whales are the largest of the toothed whales. Males can grow up to 20 metres long and can weigh up to 60 tonnes, while the females are smaller (though no less impressive). The head alone makes up one-third of its body length, and the sperm whale has the largest brain of any living animal.

Sperm whales are found throughout the world in the deep ocean, so are seldom seen near land. Female whales and their young (called calves) usually live in tropical and subtropical waters, while the males leave these breeding pods at around six years old and range further afield, as far as polar waters. This means that the population of resident and visiting sperm whales seen off the coast of Kaikōura in the South Island is almost exclusively male. They can be seen so close to shore because two very deep areas of water, the Conway Trench and the Kaikōura Canyon, run to just a few kilometres from the shoreline there.

Sperm whales produce a large amount of oil, but they are difficult to catch and kill, and can be aggressive when injured. After sticking the whale with a harpoon, whaleboats might be dragged for miles across the ocean—called a 'Nantucket sleigh ride'—or taken down with the diving whale. There are many stories of sperm whales getting the upper hand, flipping small whale-chasing boats and drowning their crew, or even attacking larger whaleships. The famous novel *Moby-Dick* by Herman Melville, published in 1851, was based on the real-life story of the American whaleship *Essex*, captained by George Pollard, which was rammed and sunk by a sperm whale in the Pacific in 1820. Cast adrift in small boats in an enormous empty ocean, the crew of twenty floated for nearly three months. Many of them died waiting to be rescued and were eaten by their surviving mates.

By the early 1850s, the sperm whale population around New Zealand was largely 'fished out' and the American whalers headed for more-lucrative waters in the northern Pacific Ocean. When New Zealand became a British colony in 1840, laws began to be enforced and duties were charged to visiting ships, so it stopped being such an attractive destination anyway. The boom days of sperm whale fishery in New Zealand were over.

Today, the sperm whale is still considered endangered. Its populations have been slow to recover from the global slaughter of the late 1700s and early 1800s. It is still found in oceans around the world, especially off Alaska in the American northwest and New England in the northeast, and around the Pacific Islands, including New Zealand. It is estimated that there are between 300,000 and 450,000 of these magnificent animals left in the world. In New Zealand, their threat classification was recently changed from 'nationally threatened' to 'data deficient'—researchers are not sure how many sperm whales are left here, and they believe that numbers may once again be declining.

Sperm whales were not the only whale to fall victim to human exploitation in the early 1800s. From the late 1820s, shore whaling stations were established targeting the southern right whale (*Eubalaena australis*). Right whales can grow up to 18 metres long—about as long as a bendy bus—and are a dark browny-black colour. They often have distinctive patches of tough white skin which may be colonised by barnacles.

Right whales were hunted for oil, rendered out of their fatty blubber, and for the baleen from their mouths—hairy plates of a hard substance called keratin, the same material that makes up human hair and fingernails. Flexible yet strong, baleen was used to reinforce seats and women's undergarments, and the hairs fringing the baleen plates were used to stuff furniture and make wigs.

Southern right whales are migratory. The females move from their summer grounds around Antarctica to warmer waters further north in winter and spring to give birth to calves. As they moved ponderously up and down the coasts of Aotearoa, the whalers killed them in their thousands.

In 1829, Jacky Guard, a freed convict from Port Jackson, was the first to set up a whaling base in New Zealand, on the edge of the Marlborough Sounds. More shore stations followed, along the east coast of the South Island as far south as Foveaux Strait, and on the other side of Cook Strait in Kāpiti, Hawke's Bay, the East Cape, Bay of Plenty and Northland. The whalers could live on shore in relative comfort, launching small boats off the beach to chase the whales that passed by. Killing and processing the whales was still a disgusting, physically demanding job, however, and whaling stations could often be smelled from some distance away.

It didn't take long for the indiscriminate slaughter to have an effect on whale populations. The slow, docile right whales (they were called this because they were the 'right' ones to kill, easy to catch and floating once dead) became scarce even in the 1840s, and the shore stations started to shut down. One of the whalers' practices was to kill whale calves first, so that the distressed mother would follow her dead offspring towards the shore where she could be killed. That way, whalers didn't have to tow the large whale so far. (This didn't work with sperm whales, apparently, which just became enraged if their calves were killed.) Pregnant whales were also targeted, as they carried more blubber and therefore produced more oil. Of course, the whalers themselves were reluctant to admit that their actions had destroyed their own industry, and maintained that the whales had simply 'moved on'. In reality, they had been hunted almost to extinction.

By 1850 right whales were all but gone from New Zealand waters. In the words of naturalist Ernst Dieffenbach, writing in 1843, the whalers had 'felled the tree to obtain the fruit', and destroyed their own industry in the process. By the early twentieth century, right whales were seldom seen—although the whalers still hunted them when they could, with thirteen taken by shore-based whaling stations between 1916 and 1926. By the 1920s, it was estimated that just 30–40 breeding females were left in New Zealand waters, around the subantarctic Auckland Islands.

Whaling wasn't over, however: after a pause, the hunters simply switched species. In the early 1900s several whaling stations were set up to pursue humpback whales (*Megaptera novaeangliae*) using motorised whale-chasers and powerful harpoon guns. Humpbacks were also targeted on their migrations around the New Zealand coast, from stations at Whangamumu just south of the Bay of Islands, the Marlborough Sounds (by the Perano whaling family), and at Great Barrier Island (Aotea). These operations continued into the 1960s, when there were too few migrating humpbacks to make hunting worthwhile. The last whale killings occurred in 1963.

The right whale was protected in New Zealand waters by the Whaling Industry Act 1935, which banned the killing of adult whales and their calves in our territorial waters. The penalty was up to three months' imprisonment and a fine of £200. A closed season was imposed on whaling of other baleen whales such as humpbacks from 1949, protecting the animals from September to April each year; then in 1964 they became totally protected. Those same regulations also introduced a protected season for sperm whales, from May to August, but later that year the whaling industry came to a natural end. The last whale taken by commercial whalers in New Zealand was a sperm whale killed by Trevor Norton from the Perano station in Tory Channel / Kura Te Au at the top of the South Island on 21 December 1964. All these pieces of legislation were repealed and the protected status of all marine mammals rolled together into the Marine Mammals Protection Act 1978, which provided complete protection for all species in New Zealand's 200-mile economic zone.

Whaling in international and Antarctic waters continued, however, with Russian and Japanese operators continuing to target humpbacks and other baleen whales. In 1994 the Southern Ocean Whale Sanctuary was established by the International Whaling Commission, and all commercial whaling is completely banned south of 40 degrees latitude.

Slowly, whales are returning to New Zealand waters. The southern right whale's threat classification was revised from 'nationally vulnerable' to 'recovering' in 2019, with the population in our waters now estimated at around 1000 animals with population growth of around 7 per cent each year. The migratory humpbacks are considered 'secure overseas', with an estimated 60,000–80,000 whales around the world.

As the right whales fight their way back from extinction, we are treated to more sightings of these beautiful animals. In August 2015, a right whale came into Auckland's Waitematā Harbour and was viewed by hundreds of people on the waterfront. In 2018, a male right whale spent a week in Wellington Harbour, leading the city to delay its Matariki fireworks display for fear of disturbing it. Another whale was spotted the following winter off Island Bay (Huritini), but this time it didn't stay around. It seems likely that right whales will become more frequent visitors as they cruise up the coasts, untroubled and finally appreciated and treasured.

The great whales are not the only marine mammals in the waters off Aotearoa. New Zealand is also home to the world's smallest dolphin—and a sub-population which is also the world's rarest. The Māui dolphin is a subspecies of Hector's dolphin (now also called the New Zealand dolphin, or *Cephalorhynchus hectori*). These endearingly cute mini-dolphins are just 1.5 metres long—about as long as a 12-year-old child is tall—with a distinctive rounded dorsal fin and a grey body with black-and-white markings around the face and underside. They are known to Māori by many names, including tūpoupou (which also means to pitch about in the water), popoto (also meaning short) and upokohue.

The dolphin's English and scientific names commemorate James Hector, the first director of the Colonial Museum (today's Te Papa). Hector was the first person to scientifically describe the dolphin in 1872, based on a specimen sent to him the year before that had been shot from a lighthouse tender (small boat) near Cape Campbell in Marlborough. Hector initially named it *Electra clancula*, or the New Zealand bottlenose (not to be confused with the actual bottlenose dolphin, *Tursiops truncatus*), and described it as 'common in Cook Strait, and on the West Coast as far south as Jacksons Bay, travelling in large schools'. The specimen was 1.3 metres long and weighed 35 kilograms. He noted, 'There is a complete skeleton and several skulls and lower jaws in the Colonial Museum, this being the most commonly cast up of any of the dolphins round the coast'.

Hector's dolphin is sadly no longer common. It is still found in patches around the west coast of both islands, and up the east coast of the South Island and lower North Island. The largest and most visible population is around Banks Peninsula, and the town of Akaroa has become a base for popular tours to see the little mammals. The southern population was reclassified from 'nationally endangered' to 'nationally vulnerable' in 2019, as new research estimated there to be a population of around 15,000 dolphins, but it still faces considerable threats from set-net fishing outside the marine reserves that have been created to protect it, as well as other human impacts such as boat strikes, tourism and undersea noise caused by searching for and mining mineral reserves.

The genetically distinct Māui dolphin (*Cephalorhynchus hectori maui*) was formally described as a subspecies in 2002. In 2006, a survey estimated there to be only around 100 Māui dolphins left alive, living in an area of the North Island west coast from Te Oneroa-a-Tōhē / Ninety Mile Beach down to Whanganui, but most commonly between the Manukau Harbour and Port Waikato, at the mouth of the Waikato River. Set-net bans were introduced in some of this range in 2003 and increased in 2008, when a West Coast North Island Marine Mammal Sanctuary was established, also placing restrictions on seismic surveying (using loud sounds to search for minerals under the sea floor) and seabed mining.

But the number of Māui dolphins continued to decline. In 2012, scientists estimated that there were just 55 left. Females only give birth once every one to four years, so natural population growth is slow—maybe only around one or two dolphins a year under ideal conditions.

Now, the population is still estimated to be fewer than 100, so Māui dolphins are ranked as being at critical risk of extinction. Scientists fear that the species will continue to decline in the coming years, despite protective measures, due to fishing activities and disease. Even with conservation efforts, there is still a serious risk that the Māui dolphin will be the next lost wonder of Aotearoa.

Kauri (*Agathis australis*)

16. The vanishing giant: Kauri

Witi and Tama stood beneath a tree fern and waited for the men they were guiding to catch their breath. These manuhiri were an unhealthy-looking bunch. Next to the strong brown limbs of Witi and Tama's people, these men had pale, pimpled skin, lank hair that faded to almost white and cheeks that turned red when they exerted themselves. And they were dirty. Oh, how they stank! The Māori men understood that these sailors had been at sea for many months, but they seemed to not care to keep their bodies clean. The visitors called themselves

French but the tangata whenua called them the Wiwi, because of the word they said over and over again when they meant āe, yes: *oui, oui.*

Many of the men were sick, and Witi and Tama's chief had allowed the leader of the Ngāti Wiwi to set up a hospital camp on Moturua Island, off Orokawa where they had their kāinga. Some of the French slept on the shore although most of them stayed on their two enormous ships—ships that were damaged and in need of repair, which was why the local men were now taking one of the visiting rangatira and some of his men into the forest.

Once the French had got their breath back, Witi and Tama led the shuffling men further into the undergrowth. The land was hilly, with steep slopes running down into thickly wooded gullies. They had to walk for a long time before they reached the grove of trees on a ridgetop, then finally:

'Titiro,' Witi said, as the rangatira panted the last few steps up to the ridge. 'Anei te kauri.'

He watched as the Frenchman tilted his head back—and back, and back—taking in the grove of pale, straight-trunked trees. Some were younger trees, only as thick as a man's leg, with smaller branches and leaves sprouting further down. But many were mature trees, giants, their trunks many, many times thicker than a man, stretching straight and smooth to erupt into a crown of leaves high in the air.

The Frenchman scratched his head, mouth agape.

Witi and Tama felt a sense of pride as he stood in awe of their mighty rākau.

'*Oui, oui*,' the visitor said, then struggled to wrap his tongue around their language. 'Ka pai.'

Aotearoa was once a land of trees. Thick forests cloaked around 80 per cent of its islands. The types of trees and plants changed with the climate, from the warm, wet subtropical north to the alpine forests of the south. The North Island was particularly heavily forested, almost entirely covered in trees. A thousand years ago, the only areas that were mainly bare of trees were the high mountains of the North Island and the Southern Alps / Kā Tiritiri o te Moana, and the dry inland plains of Central Otago.

However, the arrival of Polynesian settlers around 800 years ago saw the landscape rapidly modified. Some areas of forest had been burned by naturally occurring wildfires started by lightning, but Māori also used fire as a tool to deliberately clear land to grow food and to make hunting birds such as moa easier. The dry plains to the east of the Southern Alps were one of the main areas to be burned off, and the original forest was replaced by grasslands. Coastal areas in Hawke's Bay, Northland, Auckland and Wairarapa and parts of the Waikato and inland Bay of Plenty were also largely cleared. By 1840, when the Treaty of Waitangi was signed, an estimated

6.7 million hectares of primeval forest had already been destroyed, replaced by grassland, scrub and fern. We may never know what plants were lost from our native flora during this period.

Once European settlement began in earnest, millions more hectares of broadleaf, podocarp, conifer and beech forests fell under axe and fire. Some of the wood was logged for practical uses such as building or packaging, but much of it was simply burned to clear land for farming. Over the next 150 years, around another 8 million hectares of forest was felled, and the complex ecosystems it supported were obliterated. This act—in the name of 'progress'—resulted in the loss of not only the trees themselves, but also the birds, insects and invertebrates that lived within them. Many of the stories in this book tell of species which could not survive the destruction of their forest habitat.

Among the first forests to feel the bite of axe and saw was the mighty kauri bushland of the north. New Zealand's kauri, *Agathis australis*, is a member of an ancient group of trees spread across the South Pacific, from Malaysia and the Philippines through Papua New Guinea, Vanuatu, New Caledonia, Fiji and Australia. Kauri are slow-growing conifers (trees that reproduce through cones rather than flowers) and their lineage dates back to the age of the dinosaurs.

Kauri thrives in warm, wet subtropical climates, so in New Zealand it naturally grows only to the north of a line stretching from around the Kawhia Harbour in the west to Tauranga in the east. Its particular strongholds were Northland and the Coromandel, but kauri was once widespread over this northern portion of Te Ika-a-Māui.

In mature kauri forest, groups of giant trees stand together with minimal undergrowth. The leaves the kauri drop make the soil around them acidic, creating an area where only plants which do not compete with the trees, such as kauri grass and mingimingi, can grow. Other forest trees such as rimu, tōtara, miro, tōwai and taraire can grow between the big trees, but there is no doubt that kauri is the king, its broad crowns stretching high above the canopy.

Kauri has three distinct stages when growing: a bushy, conical form when young; then a skinny 'teenage' stage at between 30 and 50 years old, as it grows tall and sheds its lower branches (these trees were known to the early timber traders as 'rickers', a corruption of 'riggers', meaning that they were ideal for ships' rigging); and finally the mature stage, when the tree emerges above the surrounding forest canopy, its branches spreading out into a crown and its trunk thickening. Kauri can live for thousands of years and can grow up to 40 metres high—as tall as a ten-storey building—and yet one of these forest giants can be cut down in a day.

To Māori in the north, kauri was a chiefly tree (a role awarded further south to tōtara). In some traditions, its giant trunks were seen as the legs of the atua Tāne-mahuta, holding his parents, Papatūānuku (Earth) and Rangi-nui (sky), apart. The trees were largely left untouched, partly because they were revered, but also because they were so large they were impractical to fell and use. The enormous trunks of fallen kauri were sometimes hollowed out to make massive waka taua, carved out of a single piece of wood. Māori also burned the resinous gum (kāpia) which oozes from the bark to use in tattooing, or used it as a fire-starter or insect repellent. It was also a tasty, long-lasting chewing gum.

The first Europeans to record seeing kauri were the unfortunate Frenchmen of Marc-Joseph Marion du Fresne's expedition. In May 1772, two French ships sailed into Pēwhairangi, the Bay of Islands, in search of water, wood and a place for their scurvy-stricken crews to recover. Du Fresne also needed a new foremast and bowsprit for one of his ships, the *Marquis de Castries,* which had been damaged in a collision with his other ship, the *Mascarin,* on their voyage from Mauritius, an island in the Indian Ocean. The Frenchmen negotiated with local Māori to fell two trees to work into replacement spars, and a 'masting camp' was set up at the head of Manawaora Bay on the southern side of the Bay of Islands, not far from Kororāreka (Russell).

The Frenchmen based here working the felled trees into spars were lucky—they were not among those killed when du Fresne and some of his officers and men were attacked by Māori, probably in retribution for fishing in a tapu area, or as part of local tribal politics. The masting camp and the roughly shaped spars were abandoned as the French scrambled to leave the hostile shore—but not without killing several hundred Māori in retribution. The kauri timber trade had not got off to a good start.

But started it had, and in the following decades more and more ships started calling in the north in search of the straight-grained, easily workable timber from these magnificent trees. As with whaling, the first ships to carry kauri to England had transported convicts to Australia and were looking for a way to make the voyage home just as well paid. By the early 1800s, ships regularly called at the Bay of Islands and harbours further north and south, including the Coromandel Peninsula, where they dealt with local Māori to find and fell trees to meet an increasing demand for spars. Even the first missionaries to come to Aotearoa got in on the game: the Reverend Samuel Marsden took a cargo of kauri back to Sydney in 1815 to raise money to set up his little Church of England outpost at Rangihoua, on the northern side of the Bay of Islands. Local Māori were keen to trade timber, along with flax and food, in exchange for European goods such as axes, nails and guns.

One of the first substantial European settlements in the country was established at Horeke on the Hokianga Harbour—a short trip overland from the Bay of Islands. A trading post and shipbuilding yard was established there by enterprising Sydney merchants Thomas Raine and David Ramsay in 1826. When wandering artist Augustus Earle found a thriving business there the following year, he was 'greatly delighted with the appearance of order, bustle, and industry it presented'—an unexpected outpost of European civilisation in a strange land.

Following the signing of the Treaty of Waitangi and the arrival of British law in 1840, the fledgling timber industry intensified. Mills were set up around the north, and the logging and processing of kauri timber became industrialised. The huge trees had to be felled by hand, with axe and crosscut saw, but by the 1860s steam engines were being used to power many of the mills that cut the huge trunks into lengths and planks for building and boat construction. The kauri industry led to the European settlement of many parts of Northland and the Coromandel, and Māori worked in the industry as well.

There seemed to be no end to the kauri—although the millers and traders must have realised it was a finite (limited) resource. But there were just so many trees, and there was so much demand for the timber, both in New Zealand and overseas. Thousands of houses and commercial buildings were constructed of kauri timber in Auckland, around New Zealand, and in Pacific Rim cities such as Sydney and San Francisco. The beautiful wood was used for floor and wall boards, interior wall panelling and decorative details such as the 'wooden lace' seen on many Victorian-era villas. It was also made into furniture and boats, both working craft and beautiful recreational yachts and launches.

As well as the timber industry, a second wave of money could be made from the trees once the forests had been cut down. The trunks of the living trees oozed with resinous sap—but below the ground, lumps of old, solidified gum could be found. Much of it was very ancient, and it could be dug up in areas where the kauri trees were long gone. At that time, varnish for timber was made not from chemicals but from plant resins, and kauri gum proved excellent for this use. By the 1860s, huge quantities of gum was being dug up—often by Māori diggers or men from the coast of what is now Croatia, called Dalmatians—and exported for processing. The work was hard and dirty and the returns poor, but around 20,000 people were working in the industry in the 1890s.

As early as the 1860s, observers such as Austrian geologist Ferdinand von Hochstetter were warning that the kauri could be milled to extinction:

> Extensive districts . . . which formerly had been covered with Kauri woods, are now totally destitute of such; and the extermination of that noble tree progresses from year to year at such a rate that its final extinction is as certain as that of the natives of New Zealand. The European colonization threatens the existence of both, and with the last of the Maoris [*sic*] the last of the Kauris will also disappear from the earth.

In the 1870s, at the same time as voices were being raised about the loss of native bird species, Premier Julius Vogel became concerned about the rapid rate of destruction of New Zealand's native forests. He asked Parliament to protect the remaining trees, but the demand for land was too great. Australian George Perrin, appointed Conservator of Forests in several states, presented a damning report to the New Zealand Parliament on the state of the country's disappearing forests in 1897, writing:

> . . . in the face of rapid denudation of forest areas near the centres of trade and industry, the probable total extinction of the noble kauri, and the absolute certainty that thousands of acres of forest are practically perishing . . . the extraordinary fact remains that past

> governments have allowed forest conservation to retrograde [go backwards], although disaster must inevitably result from such neglect.

A few areas of forest were set aside for preservation in the early 1900s, but they were mostly odd bits and pieces. A large area of kauri forest at Waipoua, south of the Hokianga, was designated a 'state forest' in 1906, but that really just meant it was being set aside to be felled at a later date. Included in this forest is the largest-known kauri still standing: Tāne Mahuta, the Lord of the Forest, which is 51.2 metres high—half the length of a rugby field—with a trunk that is 13.8 metres round.

It was clear that both kauri gum and decent-sized trees were becoming increasingly hard to find. In 1900, only around 300,000 hectares of the 1.2 million hectares that had stood a hundred years earlier remained. Determined to make all the money they could while they could, the kauri millers stepped up production instead of backing off. The year 1906 was the peak of the industry, with 443,000 cubic metres of timber milled. But the kauri timber days were nearly over. The last few significant patches of bush were destroyed in the 1920s and 30s, including hard-to-access areas in the Coromandel and on Great Barrier Island (Aotea). At this time, much of the last of the kauri forests were simply burned off, to clear land for farming.

The government Forest Service was in charge of managing Waipoua and continued to 'selectively log' it, removing dead standing kauri but also some healthy trees. After World War II, newly formed conservation groups such as the Royal Forest and Bird Protection Society started to pressure the government to properly protect the forest. In 1952, more than 9000 hectares of kauri forest at Waipoua was protected in a special forest sanctuary. 'Selective logging' continued on the Coromandel and elsewhere in the north into the 1970s and 80s, however, and it was not until 1987, with the formation of the Department of Conservation, that all kauri on government-owned land became fully protected.

This should have been a happy ending. The logging of the great forests is over and new trees have been planted. But the truth is that today kauri is at greater risk of extinction than ever before.

In 2006, scientists noticed that some kauri in Auckland's Waitākere Ranges were dying. Two years later a new, previously unknown species of water mould was identified as the cause. *Phytophthora agathidicida,* which lives in the soil water around kauri, attacks the trees' roots and starves trees of nutrients and water. First an infected kauri's leaves will go yellow, then whole branches will wither and finally the whole tree will die. This disease, called ‘kauri dieback’, is easily spread through mould spores in the most tiny particles of soil, and it is almost always fatal to kauri trees that become infected.

So far, scientists have not found a cure for kauri dieback, but councils and government are working together to try to control its spread and reduce its impact. In some areas, such as the Waitākere Ranges, tracks have been closed and public access to areas of kauri forest has been restricted. Trees in the Waipoua and Ōmahuta forests in Northland are infected, and in 2018 trees in the Coromandel started to show signs of the disease.

It is not known what will become of the kauri. After thousands of years of dominance in the forests of the north, we may see it die out entirely during our lifetimes. Axe and saw did the initial damage, but a seemingly less aggressive human impact—the accidental spreading of a deadly disease—may be what brings about its eventual extinction. And that would be a sad day indeed.

The last word: What can I do?

The stories of the lost wonders of Aotearoa are sad ones. Many species have disappeared from these small islands in a short period of time—since human beings first arrived here around 800 years ago. None of those species has just 'died out': they have been lost to the world through the deliberate or unintentional actions of people—bringing pests, felling forests, hunting birds and mammals for food or science. Sometimes the decline went unnoticed until it was almost too late; sometimes people stood by and did nothing, or not enough, even though it was clear that special and unique creatures were on the verge of being lost to the world forever. Fortunately, as you have read, sometimes people stepped in at the last minute to stop extinctions from occurring.

We cannot judge those who came before us too harshly for their actions, which were based on different values to the ones we hold today. We can be thankful for those who did act, and we can learn from the experiences of the past and not make the same mistakes again.

It is easy to despair over what has already been lost and at the state of our environment today. We have better legal environmental protection and greater public awareness, but habitat destruction is still happening. Not all species are protected, many need greater research and more funding, and all of us face an uncertain future with climate change.

But we can't just sit back and do nothing. There are things we can all do to make sure that New Zealand's unique species have the best possible chance of surviving and thriving in the future.

Be informed. Reading this book is a good first step, and hopefully you now feel that you know more about the rare and endangered species of Aotearoa. Find out what you can about the state of New Zealand's environment and what is being done to protect it. Read books, blogs and websites (see pages 283–95), and keep up with the latest news and developments. At secondary school you may be able to choose subjects that will offer you the chance to learn more about the environment, such as biology and geography, with a view to getting involved in environmental management and conservation as a career, or doing further study.

Be involved. You are not alone—there are lots of people in Aotearoa who want to save New Zealand's natural environment and endangered species. Find out what groups are active in your area, such as Forest & Bird (forestandbird.org.nz), which also runs its Kiwi Conservation Club for young New Zealanders (kcc.org.nz). There might also be local organisations and environmental groups working on pest control and endangered-species management at a reserve or sanctuary near you (see pages 296–97). The more people who pitch in, the more these groups can achieve.

Be active. Think about what you can do in your own life to minimise your impact on the environment. You might not be able to go out into the field and work with rare species right now, but you can take action by picking up rubbish in parks and on beaches, reducing plastic use in your home, recycling, and cutting your carbon footprint by walking or cycling. Every little bit counts.

Be heard. Don't be afraid to speak up about what you believe in. People can't care about things they don't know about. Young people's voices are increasingly being heard on the political stage—just look at climate activist Greta Thunberg. Spread the conservation message and let other people know the stories of our lost wonders and how we are working to save what is left. Together, we can make a difference.

Acknowledgements

I was lucky enough to be the recipient of the 2019 New Zealand Society of Authors Auckland Museum Research Grant and Residency, which provided me with some funding, access to the museum's collections for research and, perhaps most importantly, four weeks of writing time at the Michael King Writers Centre in Devonport, Auckland. I mean it when I say that this book would not have been written without the assistance of this grant and residency, as it enabled me to dedicate time and headspace to the project. As much as we might like them to, books don't write themselves, and the weeks I spent working at the Michael King Centre were among the happiest and most productive I have spent all year! Thank you to the New Zealand Society of Authors for making this opportunity possible. Thanks also to the board and staff of the Michael King Writers Centre, especially executive director Jan McEwen, for making my stints there so enjoyable and fruitful.

At Auckland Museum, thank you to Adam Moriarty, head of collection information and access; Hugh Lilly, Julie Senior, Elizabeth Lorimer, Martin Collett and others in the library who chased up my requests and found everything I asked for; Robert Vennell, Ewen Cameron and Yumiko Baba in the botany department; Severine Hannam from marine; and Matt Rayner from natural history.

Huge thanks must also go to my publisher, Jenny Hellen of Allen & Unwin, who has been enthusiastic about this project since we first discussed it at least a decade ago. It was with her guidance and expert opinion that the book has taken its current shape, and as always I appreciate her wisdom, support and friendship. Thanks also to the rest of the Allen & Unwin team, who will sell and promote this book, and the others who have helped bring it to fruition, including Leanne McGregor and Leonie Freeman, designer Megan van Staden, editor Claire Davis and of course Phoebe Morris, whose beautiful illustrations have brought these lost wonders back to life.

I would also like to acknowledge Rhys Buckingham, for the fascinating conversation we had in which he shared his first-hand knowledge of the grey ghost, the South Island kōkako. Thanks also to Sheridan Waitai of Ngāti Kuri and Peter Bellingham of Landcare Research for the update on the Three Kings kaikōmako. I also spent a lot of time at the Research North department at Takapuna Library, researching and writing, so thank you to the Auckland Libraries staff there.

Thanks to my cohort in the University of Auckland Masters of Creative Writing programme for supporting my ongoing writing journey, especially Josie Shapiro and Erica Stretton, who regularly keep me on track. And finally, my thanks as always to my family, Rob, Florian and Natalie—without your understanding and love, none of this would be possible.

With thanks to:

KERMADEC ISLANDS
Kermadec Islan
NE of NZ
– see inset bo
NEW ZEALAND
Three Kings Island
Cape Reinga
Houhora
Bay of Islands
Ninety Mile Beach
NORTHLAND
Poor Knights Islands
Tiritiri Matangi Island
Mokohinau Island
Little Barrier Island
Great Barrier Island
Hokianga Harbour
Waipoua Forest
Waitematā Harbour
Kaipara Harbour
Auckland
Coromandel Peninsula
Manukau Harbour
Tauranga
WAIKATO
Rotorua
Taupo
New Plymouth
Pureora Forest
Te Urewera
Gisbor
Tongariro National Park
HAWKE'S BAY
Mt Taranaki
Mangahouan
Napier
Waingongoro River
Kahurangi National Park
Manu Island
Kāpiti Island
Pūkaha Mt Bruce
National Wildlife Cen
Tararua Ranges
Nelson
WAIRARAPA
Wellington
Malborough
Wairau Bar
Remutaka Ranges
Te Pokohiwi
Kaikōura
Waipara
WESTLAND
CANTERBURY
Southern Alps
Christchurch
Banks Peninsula
Chatham Islands
– see inset box
Dusky Sound /
Resolution Island
Anchor Island
Milford Sound
Mt Aspiring National Park
Burkes Pass
Timaru
Waimate
(Kapua Swamp)
CHATHAM ISLANDS
Murchison Mountains / Takahē Valley
CENTRAL OTAGO
FIORDLAND
Dunedin
SOUTHLAND
Lake Te Anau
Catlins
Codfish Island /
Whenua Hou
Stewart Island / Rakiura
Big South Cape Island

Glossary

adaptation — the process of change through which a plant or animal changes to become better suited to its environment

anthropologist — person who studies human societies and cultures and their development

archaeologist — person who studies human history and prehistory (before recorded history) through physical remains excavated from sites

artificial breeding — using technologies to help animals get pregnant and have babies

biodiversity — the variety of plant and animal life

bird of prey — bird which kills other birds or animals to feed on

botanist — person who studies plants

captive breeding — raising animals and

encouraging them to have babies in a controlled environment such as a zoo or sanctuary

conifer — tree which has cones, and often thin, needle-like leaves

cross-fostering — using one species to raise the young of another species

cryptozoologist — person who searches for and studies animals which are not proven to exist

direct-action conservation — the use of public forms of protest to promote a cause or make demands, rather than negotiation

DNA — the material present in all living organisms which carries genetic information

entomologist — person who studies insects

ethnologist — a person who studies the characteristics of human societies and the relationships between them

evolution — the process by which living organisms develop and change from earlier forms over long periods of time

excavation — digging up preserved or fossil human or animal remains, in order to study them

extinction — when the last of a species dies out, and there are no more of that type of animal or plant left in the world

fauna — animal life

feral — wild, escaped from domestication

fledging — when a baby bird develops feathers it can fly with (or mature feathers, if flightless)

flora — plant life

fossil — impressions or remains of animals or plants preserved in rock

functionally extinct — when individual plants and animals may survive but cannot breed and continue the species (e.g. if only males remain, or individuals are very far apart)

genus — scientific category that ranks above species and below family; the capitalised part of an organism's Latin name e.g. *Strigops*

geologist — person who studies the physical structure of the Earth, including rocks and their origins

germination — the sprouting of a seed, when it begins to grow into a plant

gizzard stone — rock swallowed by moa or other animals and kept in part of their stomach to help grind up and digest tough food

Gondwana — original large land mass in the southern hemisphere, off which Zealandia broke

incubation — sitting on eggs to keep them warm

Latin name — the scientifically recognised name given to an animal or plant, comprising its genus and species. Usually written in italics with the genus name capitalised e.g. *Strigops habroptilus*

lek mating — system where male animals gather to put on competitive displays to attract a mate e.g. kākāpō booming

mast year — season where trees and other plants produce a bumper crop of fruit

midden — remains of a human rubbish dump

natural history — the scientific study of animals or plants

natural selection — the process through which those animals and plants organisms which are best adapted to their environment survive better and produce more offspring, resulting in evolution of that species

naturalist — person who studies natural history nomenclature — the scientific system of naming plants and animals

ornithologist — person who studies birds and their habits

palaeontologist — person who studies fossils

pelagic — relating to the deep ocean

podocarp — a type of conifer (tree) which has fleshy-footed seeds, like little fruits

predator — animal which kills other animals for food

predator fencing — protective fence built around a sanctuary to keep out predators

remnant population — a small number of animals surviving in an area after most have died out

render — melt down blubber or fat to make oil

scavenger — animal which feeds on birds or animals which are already dead

scientific description — the initial, official description of the appearance and features of a new plant or animal

species —a group of similar plants or animals which are able to breed and produce offspring; the scientific category below genus e.g. *habroptilus*

subfossil — remains of a plant or animal which have not yet turned to rock

taxonomy — the branch of science concerned with classification, especially of animals and plants

tectonic plates — sections of the Earth's crust on which continents sit. These can move and rub together, resulting in earthquakes and volcanic activity

translocation — moving an animal or plant to a new location, to help it to survive

Zealandia — the continent which New Zealand is part of, now mostly underwater

References and further reading

I have consulted many books, journals, magazines, websites and other resources to put together the stories in this book. The following are some of the key works that have helped me on my research journey and which may be of interest if you would like to read more on the topic.

In terms of species lost in New Zealand, *Extinct Birds of New Zealand* by Alan Tennyson and Paul Martinson (Te Papa Press, 2006) is a comprehensive and attractive resource. *The Lost World of the Moa: Prehistoric Life of New Zealand* by Trevor Worthy and Richard Holdaway (Canterbury University Press, 2002) is a more detailed discussion of the birds and animals of New Zealand prior to human settlement, based on archaeological evidence. Ronald Cometti's *New Zealand Through Time: An Illustrated Journey Through 83 Million Years of Natural History* (New Holland, 2008) is another interesting illustrated book.

Nineteenth and early twentieth-century observers I have read and quoted include: Walter Buller, whose *A History of the Birds of New Zealand* (1888 edition) is available online through the New Zealand Electronic Text Collection—Te Pūhikotuhi o Aotearoa; Charles Douglas, 'Mr Explorer Douglas', whose writings can be found in *Mr Explorer Douglas: John Pascoe's New Zealand Classic*, edited by John Pascoe and revised by Graham Langton (Canterbury University Press, 2000); and Andreas Reischek (there are several editions of his translated *Yesterdays in Maoriland: New Zealand in the 'eighties*). Ross Galbreath's biography *Walter Buller: The Reluctant Conservationist* (GP Books, 1989) is another interesting insight into the naturalist and collector.

Working for Wildlife: A History of the New Zealand Wildlife Service, also by Ross Galbreath (Bridget Williams Books, 1993), gives a good overview of the development of the government conservation service in New Zealand, including the stories of the takahē, kākāpō and black robin. *Our Islands, Our Selves: A History of Conservation in New Zealand* by David Young (University of Otago Press, 2004) is another useful summary. *Wild South: Saving New Zealand's Endangered Birds* by Rod Morris and Hal Smith (TVNZ/Century Hutchinson, 1988) also tells first-hand the stories of some of our great conservation campaigns.

For Māori legends and traditional uses of birds, I recommend Murdoch Riley's *Māori Bird Lore: An Introduction* (Viking Sevenseas, 2001) and *Manu Māori: Bird Legends and Customs* (Viking Sevenseas, 2006).

The well-researched and informative website nzbirdsonline.org.nz is a project led by Te Papa curator Colin Miskelly. It contains a wealth of information on the birds of Aotearoa, both extinct and surviving. The excellent and comprehensive Te Ara—The Encyclopedia of New Zealand website (teara.govt.nz) and its Dictionary of New Zealand Biography (teara.govt.nz/en/biographies) have been vital resources. The Department of Conservation (DOC) website (doc.govt.nz) provides lots of information about our endangered species and the efforts being made to save them. New Zealand Geographic magazine (nzgeo.com) has many excellent articles available online about our extinct species and conservation efforts today.

In terms of scientific articles, I am very grateful that the *Transactions and Proceedings of the Royal Society of New Zealand* from 1868 to 1961 (rsnz.natlib.govt.nz/index.html), and the journal of the Ornithological Society of New Zealand, *Notornis,* (notornis.osnz.org.nz/publications) are available online. The Biodiversity Heritage Library (biodiversitylibrary.org) is an international initiative to create the world's largest open-access digital library of works about life on Earth, and I was able to access a number of historical publications through its portal. For local topics, Papers Past (paperspast.natlib.govt.nz), a searchable database of historical newspapers and other publications from around New Zealand, is another fun and fascinating resource.

In terms of extinction on a wider scale, my inspiration for this work was a book I read when I was about ten years old, called *The Dodo, the Auk, and the Oryx: Vanished and Vanishing Creatures,* by Robert Silverberg (Puffin, 1967). This book told the extinction stories of many species around the world, including the moa, and made me want to tell the stories of species that had been lost from my own country. It's still a great read, and you might be able to find it in a library or second-hand bookstore. *A Gap in Nature: Discovering the World's Extinct Animals* by Tim Flannery and Peter Schouten (Text, 2001) is a beautifully illustrated book featuring the stories of more than 100 species around the world which have become extinct in modern times.

The following are some subject-specific resources I used and other books and websites which may be of interest.

1. What came before: New Zealand's dinosaurs

Candler, Gillian and Ned Barraud, *From Moa to Dinosaurs: Explore & Discover Ancient New Zealand*, Potton & Burton, 2016

Cox, Geoffrey and Joan Wiffen, *Dinosaur New Zealand*, HarperCollins, 2002

Hill, David, *Dinosaur Hunter: Joan Wiffen's Awesome Fossil Discoveries*, Picture Puffin, 2019 (illustrated by this book's illustrator, Phoebe Morris)

Mortimer, Nick and Hamish Campbell, *Zealandia: Our Continent Revealed*, Penguin, 2014

'New Zealand Fossils', GNS Science Te Pū Ao, www.gns.cri.nz/Home/Learning/Science-Topics/Fossils/NZ-fossils

Stevens, Graeme, Matt McGlone and Beverley McCulloch, *Prehistoric New Zealand*, revised edition, Reed, 1995

Wiffen, Joan, *Valley of the Dragons: The Story of New Zealand's Dinosaur Woman*, Random Century, 1991

2. The biggest birds: Moa

Anderson, Atholl, *Prodigious Birds: Moas and Moa-hunting in Prehistoric New Zealand*, Cambridge University Press, 1989

Berentson, Quinn, *Moa: The Life and Death of New Zealand's Legendary Bird*, Potton & Burton, 2012

Duff, Roger, *Pyramid Valley*, Pegasus Press for the Association of Friends of the Canterbury Museum, 1952

Eyles, James R., *Wairau Bar Moa Hunter: The Jim Eyles Story*, River Press, 2007

Holdaway, Richard and Trevor Worthy, 'Lost in Time', *New Zealand Geographic*, issue 12, Oct–Dec 1991, www.nzgeo.com/stories/lost-in-time

Wolfe, Richard, *Moa: The Dramatic Story of the Discovery of a Giant Bird*, Penguin, 2003

Worthy, Trevor and Richard Holdaway, *The Lost World of the Moa: Prehistoric Life of New Zealand*, Canterbury University Press, 2002

3. Terror in the skies: Haast's eagle

Greenaway, Richard L. N., 'Frederick Richardson Fuller 1830–1876', *Rich Man, Poor Man, Environmentalist, Thief*, Christchurch City Libraries, 2000, www.christchurchcitylibraries.com/heritage/publications/richmanpoorman/frederickrichardsonfuller

Holdaway, Richard, 'Terror of the Forest', *New Zealand Geographic*, issue 4, Oct–Dec 1989, www.nzgeo.com/stories/terror-of-the-forest

Worthy, Trevor and Richard Holdaway, *The Lost World of the Moa: Prehistoric Life of New Zealand*, Canterbury University Press, 2002

4. New killers take their toll: Whēkau, koreke, piopio

Pascoe, John (ed.), *Mr Explorer Douglas: John Pascoe's New Zealand Classic*, revised by Graham Langton, Canterbury University Press, 2000

Worthy, Trevor and Richard Holdaway, 'Laughter in the Night', *New Zealand Geographic*, issue 32, Oct–Dec 1996, www.nzgeo.com/stories/laughter-in-the-night

5. The dying song: Huia

Galbreath, Ross, *Walter Buller: The Reluctant Conservationist*, GP Books, 1989

Phillipps, W. J., *The Book of the Huia*, Whitcombe & Tombs, 1963

Szabo, Michael, 'Huia, the Sacred Bird', *New Zealand Geographic*, issue 20, Oct–Dec 1993, www.nzgeo.com/stories/huia-the-sacred-bird

6. The light that went out: Lyall's wren

Brown, Derek, *Stephens Island: Ark of the Light*, D. Brown, 2000

Ure, Graeme, 'Island refuge', *New Zealand Geographic*, issue 32, Oct–Dec 1996, www.nzgeo.com/stories/island-refuge

7. In search of the grey ghost: South Island kōkako

Bensemann, Paul, *Fight for the Forests: The Pivotal Campaigns that Saved New Zealand's Native Forests*, Potton & Burton, 2018

Evans, Kate, 'In Search of the Grey Ghost', *New Zealand Geographic*, issue 140, July–Aug 2016, www.nzgeo.com/stories/in-search-of-the-grey-ghost

Marsh, Sid and Brian Parkinson, 'Kokako', *New Zealand Geographic*, issue 87, Sept–Oct 2007, www.nzgeo.com/stories/kokako

Pascoe, John (ed.), *Mr Explorer Douglas: John Pascoe's New Zealand Classic*, revised by Graham Langton, Canterbury University Press, 2000

South Island Kōkako Charitable Trust, www.southislandkokako.org

8. The tragedy of Big South Cape: South Island snipe, Stead's bush wren, greater short-tailed bat

Ballance, Alison, *Don Merton: The Man Who Saved the Black Robin*, Reed Books, 2007

Galbreath, Ross, *Working for Wildlife: A History of the New Zealand Wildlife Service*, Bridget Williams Books, 1993

Morris, Rod and Hal Smith, *Wild South: Saving New Zealand's Endangered Birds*, TVNZ/Century Hutchinson, 1988

Wild South: Island eaten by rats, 1988, www.nzgeo.com/video/island-eaten-by-rats

9. No island refuge: The strange—and vanished—species of the Chatham Islands

Ballance, Alison, *Don Merton: The Man Who Saved the Black Robin*, Reed Books, 2007

Butler, David and Don Merton, *The Black Robin: Saving the World's Most Endangered Bird*, Oxford University Press, 1992

Chatham Island Taiko Trust, www.taiko.org.nz
Jones, Jenny, *Black Robin*, Heinemann Education, 1994
Miskelly, Colin (ed.), *Chatham Islands: Heritage and Conservation*, Canterbury University Press, 2008

10. Plenty more fish in the sea: Grayling

McQueen, Stella, *A Photographic Guide to Freshwater Fishes of New Zealand*, New Holland, 2013
Walrond, Carl, 'Out of the Frying Pan: Into Oblivion', *New Zealand Geographic*, issue 75, Sept–Oct 2005, www.nzgeo.com/stories/out-of-the-frying-pan-into-oblivion

11. The kings of hide and seek: Takahē and moho

Galbreath, Ross, *Working for Wildlife: A History of the New Zealand Wildlife Service*, Bridget Williams Books, 1993
Grzelewski, Derek, 'Takahe—The Bird that Came Back from the Dead', *New Zealand Geographic*, issue 41, Jan–March 1999, www.nzgeo.com/stories/takahe-the-bird-that-came-back-from-the-dead
Jones, Jenny, *Takahe*, Reed, 2005
Lee, W. G. and I. G. Jamieson (eds.), *The Takahe: Fifty Years of Conservation Management and Research*, University of Otago Press, 2001
Morris, Rod and Hal Smith, *Wild South: Saving New Zealand's Endangered Birds*, TVNZ/Century Hutchinson, 1988

12. The rarest plants in the world: *Tecomanthe speciosa, Pennantia baylisiana*

Judd, Warren, 'The Clifftop World of the Three Kings', *New Zealand Geographic*, issue 29, Jan–March 1996, www.nzgeo.com/stories/the-clifftop-world-of-the-three-kings

'Pennantia baylisiana', New Zealand Plant Conservation Network, www.nzpcn.org.nz/flora_details.aspx?ID=32

'Pennantia baylisiana (Three Kings Kaikomako)', Citizen Science Hub for Taranaki, www.citscihub.nz/Phil_Bendle_Collection:Pennantia_baylisiana_(Three_Kings_Kaikomako)

'Tecomanthe speciosa', New Zealand Plant Conservation Network, www.nzpcn.org.nz/flora_details.aspx?ID=41

'Tecomanthe speciosa (Three Kings Vine)', Citizen Science Hub for Taranaki, www.citscihub.nz/Phil_Bendle_Collection:Tecomanthe_speciosa_(Three_Kings_Vine)

13. All the small things: Vanished lizards, frogs, worms and insects

Bowie, Mike, *The Canterbury Knobbled Weevil: Conservation Guide*, Lincoln University, 2009, livingheritage.lincoln.ac.nz/nodes/view/2648

Dickison, Mike, 'What Happened on Stack H?', *New Zealand Geographic*, issue 159, Sept–Oct 2019, www.nzgeo.com/stories/what-happened-on-stack-h

'Ten Most Endangered', Endangered Species Foundation, www.endangeredspecies.org.nz/projects/10-most-endangered (includes fact sheets on the Canterbury knobbled weevil and Mokohinau stag beetle)

14. Parrot of the night: Kākāpō

Ballance, Alison, *Kākāpō: Rescued from the Brink of Extinction*, Potton & Burton, 2018

Butler, David, *Quest for the Kakapo*, Heinemann Reed, 1989

Ell, Sarah, *Sirocco the Rock-Star Kakapo*, Random House, 2012

Galbreath, Ross, *Working for Wildlife: A History of the New Zealand Wildlife Service*, Bridget Williams Books, 1993

Jones, Jenny, *The Kakapo*, Reed, 2003

Montgomery, Sy, *Kakapo Rescue: Saving the World's Strangest Parrot*, Houghton Mifflin Books for Children, 2010

Morris, Rod and Hal Smith, *Wild South: Saving New Zealand's Endangered Birds*, TVNZ/Century Hutchinson, 1988

White, Rebekah, 'Decoding Kākāpō', *New Zealand Geographic*, issue 141, Sept–Oct 2016, www.nzgeo.com/stories/decoding-kakapo

15. Te whānau puha: The great whales and the Māui dolphin

Ell, Sarah, *There She Blows: Sealing and Whaling Days in New Zealand*, Bush Press, 1995

Frankham, James, 'Deep Trouble', *New Zealand Geographic*, issue 110, July–Aug 2011, www.nzgeo.com/stories/mauis-dolphin-deep-trouble

Morton, Henry, *The Whale's Wake*, University of Otago Press, 1982

Slooten, Liz and Steve Dawson, *Dolphins Down Under: Understanding the New Zealand Dolphin*, Otago University Press, 2013

Wakefield, Edward Jerningham, *Adventure in New Zealand, from 1839 to 1844: with some account of the beginning of the British colonization of the islands*, Wilson and Horton, 1971 (facsimile)

16. The vanishing giant: Kauri

The Kauri Museum, Matakohe, kaurimuseum.com

Keep Kauri Standing, Kia Toitu He Kauri, kauridieback.co.nz

Orwin, Joanna, *Kauri: Witness to a Nation's History*, 2nd edition, New Holland, 2019

Stewart, Keith, *Kauri*, Penguin/Viking, 2008

Conservation organisations and other websites

Conservation Volunteers NZ: www.conservationvolunteers.co.nz

Department of Conservation: www.doc.govt.nz/get-involved; Kiwi Guardians programme: www.doc.govt.nz/parks-and-recreation/places-to-go/toyota-kiwi-guardians

Endangered Species Foundation: www.endangeredspecies.org.nz

Forest and Bird: www.forestandbird.org.nz; Forest and Bird Youth: www.forestandbird.org.nz/our-community/forest-bird-youth; Kiwi Conservation Club: www.kcc.org.nz

Green Party of Aotearoa New Zealand: www.greens.org.nz; Young Greens: www.younggreens.org.nz

Keep New Zealand Beautiful: www.knzb.org.nz

Ministry for the Environment: www.mfe.govt.nz

Nature Space: www.naturespace.org.nz (list of community conservation projects)

Ornithological Society of New Zealand: www.birdsnz.org.nz; Young Birders New Zealand: www.youngbirdersnz.com

Predator Free NZ: www.predatorfreenz.org

Sanctuaries of New Zealand: www.sanctuariesnz.org

Science Learning Hub: www.sciencelearn.org.nz

Wild Eyes: www.wildeyes.co.nz (nature missions for Kiwi kids)

World Wildlife Fund New Zealand: www.wwf.org.nz

Index

Q

R

S

T

Sarah Ell was born and raised on Auckland's North Shore and has always had a close relationship with nature and the outdoors. Sarah trained as a newspaper journalist before working in magazines and book publishing. She is the author of 10 books, several of them for young people, including *Sirocco, the Rock Star Kakapo* (2012). She still lives on the North Shore, with her husband, yacht designer Rob Shaw, and their two children.